WELL PLAYED

Building Mathematical Thinking Through Number and Algebraic Games and Puzzles

GRADES 6–8

Linda Dacey, Karen Gartland,
and Jayne Bamford Lynch

www.stenhouse.com

Portsmouth, New Hampshire

Stenhouse Publishers
www.stenhouse.com

Library of Congress Cataloging-in-Publication Data

Dacey, Linda Schulman, 1949-

Well played: building mathematical thinking through number and algebraic games and puzzles, grades 6–8 / Linda Dacey, Karen Gartland, and Jayne Bamford Lynch.

pages cm

Includes bibliographical references.
ISBN 978-1-62531-034-7 (pbk. : alk. paper) -- ISBN 978-1-62531-077-4 (ebook) 1. Arithmetic--Study and teaching (Primary) 2. Counting. 3. Mathematics--Study and teaching (Primary) I. Gartland, Karen. II. Lynch, Jayne Bamford. III. Title.

QA135.6.D3325 2016

372.7'044--dc23

2015022556

Cover design by Alessandra S. Turati

Interior design and typesetting by Victory Productions, Inc.

Printed in the United States of America

This book is printed on paper certified by third-party standards for sustainably managed forestry.

26 25 24 23 22 4371 9 8 7 6 5 4 3

ONTENTS

Games and **Puzzles** Listed in Alphabetical Order

ACKNOWLEDGMENTS

We are deeply indebted to the teachers and students who collaborated with us during the development of this project. Each game was explored within a classroom, and teacher and student insights permeate this book. We are particularly grateful to students for serving as our consumer experts. Their feedback helped us fine-tune our thinking and play more with ways to embed key mathematical ideas into our discussions of the games and puzzles.

We are grateful to everyone at Stenhouse, but most particularly Toby Gordon, who showed such early faith in us. Her words "Write about what matters most to you" gave us the freedom to explore, reflect, play, and puzzle. And then, of course, she gave us such valuable and timely feedback all along the way. We are also grateful to our outstanding outside reviewer, who probed our thinking with important insights and intriguing questions. Thank you also to Chris Downey, Louisa Irele, Stephanie Roberts, and Jay Kilburn for their care and expertise. All of you added greatly to the quality of our work.

Introduction

It is a happy talent to know how to play.

—Ralph Waldo Emerson

Our whole life is solving puzzles.

—Erno Rubik

Did you have a favorite game or puzzle as a child? Why did you like it? Looking back, what do you think you learned from it?

People have engaged in playing games and solving puzzles for thousands of years. Games and puzzles continue to provide important opportunities for children to experience playful learning. And, as the proliferation of online game playing and puzzle solving shows, these activities continue to capture our interest.

Games and puzzles based on logical thinking are often linked to mathematics. Out of school, they are considered recreational. In school, games and puzzles often provide opportunities for students to practice skills. We see their incorporation in math lessons more than in other subject areas. Within mathematics, they tend to focus on computation, with the goal of increasing fluency. Games and puzzles are included in many mathematics curriculum resources, and teachers might offer them as a choice, as an independent activity, or as a rotation during instructional time.

Why This Book?

So, with interest in games and puzzles fully established and lots of games available to teachers through online sources and curriculum materials, why did we want to write this book? As a way to begin to answer this question, we'd like to share something we witnessed in an eighth-grade classroom when we asked a teacher to have her students play *A-Mazing* (see page 40), a game we created, to learn more about how mathematical games were explored at the middle school level. The game requires players to take turns rolling a die and moving that many places on the game board. The move may be horizontal, vertical, diagonal, or any combination of those directions. Each number of the game board passed over during the move is then used in an equation, which is recorded. When one team reaches the end of the board, each team finds the sum of all its equations. The team with the lesser sum wins.

The teacher reviewed the directions of the game and assigned teams. One team created the expression $\frac{1}{3} \times 1.2 - 5^2 + \frac{3}{8}$. Finley said, "Okay, so we multiply and then just do the subtraction and addition." Nadia said, "No, we have to do the exponent first because that's the order of operations rule." They checked with their opponents, who told them it didn't matter as long as they squared the five before they subtracted. Both Nadia and Finley's facial expressions suggested that they were surprised by this response. Nadia said, "It has to matter." Finley then nodded and quietly suggested, "Let's just use different cards, without exponents, and not be wrong." As a result of this decision, all of Nadia and Finley's recorded equations were correct, but they were unable to resolve their partial understanding while playing the game.

We know it is impossible to monitor all students all of the time, and this teacher had checked in with these players during the game and noted that their recordings were correct. However, observing this interaction caused us to think more about the use of games and puzzles in our classrooms. We began with the question *How could we increase the likelihood that this lack of understanding would be noticed and addressed?* We wondered how best to support students in taking responsibility for ensuring that understanding is reached, and how to help teachers focus their quick assessments of student work. Our conversation then moved quickly to a general discussion about games and puzzles and how we might increase their potential for deepening students' conceptual learning as well as their computational fluency. After all, middle school is "a time for students to make sense of mathematics, build a solid foundation and enthusiasm, and set the course for the highest level of mathematics in the future" (Tassell, Stobaugh, and Sheffield 2011, 1).

It was clear that we wanted to consider the use of games in middle school classrooms more fully. We began our work by identifying our favorite games and puzzles and asking teachers to do the same. This process led us to adapt some games and puzzles as well as create new ones. Then we tried to think about how to deepen students' learning through their exploration of games and puzzles. We asked ourselves key questions, including

- Why is this game or puzzle worth exploring?
- How could student-to-student math talk be increased?
- How could teachers support learning as students played and solved?
- What might teachers notice as students played the game or solved the puzzle that would inform future instruction?
- What assessment tasks could reinforce student accountability?
- What task would provide an opportunity to extend students' thinking?

Is It a Game or a Puzzle or an Activity?

One of the surprises of this work was how murky the distinctions can be among games, puzzles, and tasks. Is a computer game that requires a player to find clues and the correct path to reach a certain goal a game or a series of puzzles? Are we playing a game when we solve a puzzle? Is categorizing numbers by their attributes a puzzle, a task, or a game? Is an activity a game if students take turns, and a puzzle if it uses clues? Koster suggests that "games are puzzles to solve, just like everything else we encounter in life" (2013, 34). Note that both games and puzzles

- involve sequencing and pattern recognition;
- require strategy; and
- offer competition against an opponent, or the clock, or your own abilities to reach a solution.

There are, of course, some differences as well. For example, puzzles can be lost only by giving up.

We have identified the games and puzzles in this book as either one or the other, but we found the following criteria important to both:

- It addresses important mathematical ideas.
- It is engaging.
- It offers a range of difficulty levels.
- It requires and stimulates mathematical insight.
- It supports the habits of mind essential for success with mathematics and real-world problem solving.

How Is This Book Organized?

Chapter 2 addresses instructional decisions related to games and puzzles in the classroom. Our goal in the chapter is to address ways to support teacher orchestration of gaming and puzzling and students' involvement in the process. We pay particular attention to helping students take responsibility for their learning of mathematics.

The next five chapters of this book focus on content-specific games and puzzles arranged by content focus: number systems; ratio and proportionality; expressions and equations; statistics and probability; and functions. Please remember, though, that there is a lot of overlap among these topics and sometimes, a student's strategy may be related to multiple chapters. There are five fully developed games or puzzles within each chapter as well as a section that suggests online games and puzzles (including apps) that are appropriate for the classroom. The online section is less detailed, as such resources change frequently. Within this section, we identify those electronic resources that are free.

The discussion of each of the five games and puzzles is organized to address the goals we identified when we began this project. Along with the expected sections "Math Focus" and "Directions," each discussion includes a section called "How It Looks in the Classroom," which shares a brief classroom story from our field-testing. "Tips from the Classroom" provides further ideas for supporting classroom implementation, some of which came from our student field-testers. "What to Look For" identifies key ideas and misunderstandings that our experience suggests will be tapped, allowing you to collect data to inform instructional decisions and note student growth over time. The "Variations" section suggests ways to change the game or puzzle to better reach the range of learners in your class, to sustain its worthiness as student learning progresses, or to adapt it to better fit another grade level than that indicated in the classroom vignette. Nearly all of the games and puzzles can be adapted to sixth- through eighth-grade students and the suggested variations will help you make such changes. The section "Exit Card Choices" provides some suggested questions you can pose to students after their experience with a game or puzzle. These tasks serve as a way to bring closure to the initial experience, while assessing students' content knowledge. Such questions reinforce students' accountability for their own learning. Responses can be individual or, as appropriate, completed in teams, and can inform your instructional decision making.

As teachers, we recognize the value in partnering with students about their learning. The more we communicate to students the role that games and puzzles play in supporting their understanding of key mathematical concepts and the use of mathematics in the real world, the better. Recording sheets and exit questions allow students to share their strategies and knowledge, and provide teachers with the opportunity to assess learning. At the end of each game or puzzle discussion, the "Extension" section, as you would expect, gives you an idea for extending the learning. As you gain familiarity, you, or your students, may create other examples of such questions and tasks.

No doubt you may be familiar with some of the games and puzzles included in this resource; you have your own favorites, too. Nonetheless, we are confident that you will appreciate the opportunity to think more deeply about the use of games and puzzles in the classroom and find new ways to make their exploration more effective for engaging students and deepening mathematical understanding.

Supporting Learning Through Games and Puzzles

As students enter their seventh-grade classroom they read the agenda written on the board, which tells them to form a group of three and take one copy of the puzzle sheet in the folder hanging next to the agenda. The puzzle provides students with clues they must use to identify what each person bought at a store. The directions explain that the students are to read the clues together and keep track of their thinking about the clues, with each solver using a different color pen or pencil. Students form groups easily; some look for particular partners, while others gather with those who enter the classroom at the same time as they do. They collect the materials needed and begin.

Later, as the teacher circulates, she overhears Meg say, "Wait, we have a lot of blue and red ink here but not so much green." Eli pauses a moment and then says, "I'm not totally following you guys. Could we stop a minute so you could catch me up?"

As simple as this scene may appear to an outside observer, teachers know that it reflects deep values related to students sharing responsibility for their learning. Teachers will also recognize the instilling of routines and setting of expectations that must precede such behavior. Thoughtful orchestration of mathematical games and puzzles goes well beyond offering opportunities to play and solve. We need to think about when and how to introduce games and puzzles and when to make them available for small-group or individual use. We must help students understand the purposes of and expectations for playing and solving and look for ways to share the responsibility of the learning process with them.

We want to set and assess learning goals, support math talk, and meet the needs of individual learners while pursuing such activities. We should recognize ways to organize students and materials to support success, and we want to involve families and caregivers in the playing and solving.

Using Games and Puzzles in the Classroom

We believe that the instructional potential of games and puzzles could be greater realized in most middle school classrooms. Too often we've seen them provided as activities with little to no follow-up, or offered to students as choices only after they have mastered the related mathematical content. We've seen that teachers who implement good teaching practices, such as asking significant debriefing questions after a problem-solving experience, often fall short of utilizing such practices with games and puzzles. Further, the games and puzzles used most frequently in classrooms tend to develop only procedural expertise, without attending to conceptual understanding. As a result, many students experience games or puzzles as fun activities or time fillers, but do not consider them as essential to their learning or as an important part of a lesson for which they are accountable. If we are to use games and puzzles to develop conceptual understanding as well as to build fluency, we must re-evaluate how we are using them and be willing to make some changes in our classroom practice.

Changing Purposes of Games over Time

The same game or puzzle can serve many different purposes, and those purposes change with increased exposure. As summarized in Figure 2.1, we view the *introduction* stage of a game or puzzle as an experience that exposes students to different ways of thinking and piques their interest. This introduction may be the focus or a component of the day's lesson. Students have several opportunities for follow-up play in teams during the *exploration* stage. During this exploration, students' learning deepens, and some generalizations are formed. Students are actively involved with initial large-group discussions and engage in conversations with peers throughout this stage. During the *variation* stage, changes support continued learning, and as a result, the interest level in the game and its appropriateness for conceptual development are maintained for a longer period of time. Subsequent play or puzzling may be intended for continued practice, rather than for the development of ideas. This *practice* stage supports automaticity when such reinforcement is needed, and students often prefer such an activity to a traditional worksheet. At this stage, learners are more likely to be working alone or playing against a single opponent. Sometimes games continue to be played as favorites, long after they have met the goal of supporting the development of conceptual understanding or computational fluency. When this *recreation* stage is reached, we encourage you to have students play the game when there is class time set aside for review or during an after-school program, allowing your class to investigate other mathematical concepts in the limited instructional time available.

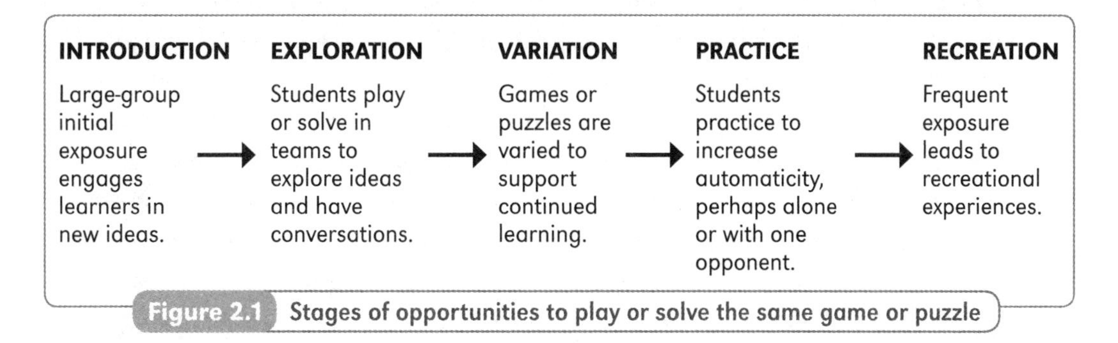

Figure 2.1 Stages of opportunities to play or solve the same game or puzzle

Building Student Responsibility for Learning

According to Alison Cook-Sather (2010), students deemed responsible are often those who do what adults tell them to do and learn the material that adults offer them. That is not our view of student responsibility. We believe that for students to become engaged, life-long learners, they need to share responsibility for their learning, from the beginning. Shared ownership is particularly important when the purposes of playing or solving include opportunities to develop conceptual understanding. Learning while playing games and solving puzzles requires clear expectations for accountability, for perseverance when the content is considered challenging, and for appropriate game-playing and puzzle-solving behavior. In traditional instructional settings, students may be more likely to assume that they will be held accountable for learning; but they may not hold that expectation for playing games and solving puzzles. In the following reflection, an eighth-grade teacher shares her thoughts about introducing games and puzzles in the classroom and holding students accountable for learning.

TEACHER REFLECTION

I was excited to introduce games and puzzles as an instructional strategy. A summer workshop helped me understand that such activities were for more than just extra practice or for fun on a Friday after students were done with the in-class activity. I waited until the third week of school, as I wanted to know the students a bit first and be sure they understood my expectations in the classroom.

I introduced a game to my students and went over the directions with them. I assigned teams and identified areas in the room for them to play. They found

their assigned partners and spaces readily, but that was the last thing that went as I expected. One group spent a long time arguing about which team went first, and another group wanted to change the rules in the middle of the game so that they could win. Others seemed to be playing more casually than I had expected and didn't make joint decisions with their teammates.

I introduce games and puzzles differently now. Rather than just going over the directions, we play a few rounds as a class or solve a similar, but simpler, puzzle. Taking time to discuss the mathematics involved seems to help students recognize that the game or puzzle is a learning opportunity. We also hold discussions about what students talked about in their game groups or with their puzzle partners, and how these conversations helped them further conceptualize what they were learning. Students often talk about how they were frustrated when they didn't understand a problem that was posed in a game or puzzle, and we discuss how to work through the challenge. Over time, their conversations changed to how they worked through challenges with the help of teammates.

I also started using exit cards, which helped students understand that they would be held accountable for learning the mathematics involved in the game or puzzle. I often chose several of the responses to discuss with the class and, as a result, student learning was extended even further. One time recently I asked them some questions on a unit test that related to the game they had played that week.

I was right about the value of games and puzzles in the classroom. I just had to help my students see that value and set expectations for their learning. Many of my students still bring competitive energy to these activities, but they also bring concentration and perseverance. Both my students and I have a better appreciation for what it means to learn through playing and puzzling.

As suggested in this teacher's reflection, playing games and solving puzzles provide an engaging context in which students can learn to work through their challenges and struggles with mathematics content.

Encouraging Perseverance

The research from Dr. Carol Dweck, author of the book *Mindset* (2007), reveals that students who are invested in their learning and recognize the effort that it takes to learn something new have a growth mindset; they feel encouraged to work harder to learn more. As playing games and doing puzzles usually requires the willingness to accept that you won't always win or that you won't always get it right the first time, students are given the opportunity to practice persistence and productive struggle—key elements of a growth mindset.

So, how do we teach students to believe that persistent effort is worth it; that they shouldn't just give up? Hiroko Warshauer (2015) suggests four teaching strategies that support students in their productive struggle:

- Ask questions that focus students on their thinking.
- Encourage student reflection and support student effort.
- Give students time to manage their own struggles.
- Acknowledge the important role struggle plays in learning.

Our conversations with middle school students suggest that they are often persistent when they are learning how to play a new sport or instrument. They know that they have to practice many times over to hit a baseball the way they want to, or to play the same song for weeks in order to achieve mastery. Encouraging students to relate those experiences to playing games and solving puzzles helps them to push through a difficult problem or a misunderstanding, because they know that it will allow them to improve.

Wilburne and her coauthors (2014) suggest that it is not enough to just talk about perseverance. Some of the teachers we interviewed about how they helped their students learn more about perseverance proposed having them rate, on a scale of 1 to 5, how much they persevered when they played a game or solved a puzzle. Other teachers suggested having students write about what helped them or prevented them from persevering. You could have your students write similarly, as a reflection on how they persevered when they played or solved. Thoughts provided by an eighth-grade student are shown in Figure 2.2.

When I was playing the game at first
we wanted to give up because it was hard
but we kept going. I knew that even if I had
made a mistake that I could learn from it the next
time. Working together helped me persevere more then I would
have if I was playing the game myself.

Figure 2.2 Written reflection about perseverance

Clarifying Expectations for Playing Games and Solving Puzzles

Some teachers view the setting of expectations and the sharing of responsibilities as separate topics, but we see them as intertwined. Students cannot be responsible without a clear understanding of what is expected of them. We also think students need to share in identifying those expectations. Further, students should have a variety of opportunities to learn ways to demonstrate positive behaviors related to games and puzzles.

Protocols

Working with students to create game and puzzle protocols, rules, or norms they should all follow when playing or solving helps establish and clarify expectations. Such protocols should be posted in the classroom, where they are visible to all, and be referenced before, during, and after gaming and puzzling.

After a seventh-grade teacher talked with her son about playing games online, she thought about how often gamers play anonymously and how such experiences might impact students' conduct when they are asked to play games together in a classroom. She decided that she wanted her students to talk about behavioral expectations in relation to games in the math classroom. She knew her students were familiar with protocols from their science lab experiences and she decided they could create one for playing and puzzling to learn mathematics. A rich discussion ensued about rules and expectations. A number of students chimed in when they discussed how to take seriously the work of writing on a recording sheet. Some thought that they would do it only if they were going to be graded. Others said they understood that it was there to help them remember what they did when they were playing or solving. Collectively, they landed on a statement that all could agree to. The teacher and the students were pleased with their final protocol and agreed to adhere to it as they played and puzzled. (See Figure 2.3.)

When we play games and solve puzzles, we will:

- Take time to talk about the math we are learning.
- Help our teammates to understand, learn, and stay on task.
- Give mathematical arguments for our choices and challenge each other's thinking until we reach agreement.
- Be respectful of each other and the materials.
- Keep the conversation focused on the math.
- If there is a recording sheet, communicate our thinking accurately.
- Put forth our best effort and be good sports.

Figure 2.3 Math games protocol

Gaming and Solving Etiquette

Your students are likely familiar with online etiquette; for instance, they may know not to type in all caps because that is considered screaming when they write an e-mail or send a text. They also have ideas as to how to treat one another respectfully in the classroom community. But along with a protocol, it can be helpful to discuss specific situations that arise in relation to playing and solving, particularly when doing so in teams.

In response to teachers asking us about ways to help students apply expectations for "classroom citizens" to game and puzzle situations, we created a hypothetical "game-and-puzzle etiquette expert" who would give advice about how to deal with difficult situations that might arise. We asked students to take on the role of this expert and developed scenarios for them to consider. We found that with some adaptations, these scenarios worked across a variety of grade levels. When a sixth-grade teacher asked us about ways to help students be better team members, we suggested some scenario cards, and he agreed to try this technique with his students.

He had the students sit in groups and send a member to randomly choose one of the cards shown in Figure 2.4, which he had spread out facedown. (A copy of these cards can also be found on page A-3.) He told them that they had been hired as game-and-puzzle etiquette experts and their job was to discuss the concern raised, create a written response to the writer's dilemma, and identify one "expert" in their group who would share it with the class. After listening to each etiquette expert's response, the whole group discussed several of their ideas for how the individual could make the situation better. For example, the students listed the following suggestions for what a student could do if a friend were fooling around too much and not concentrating while they were trying to play a game: remind the friend that they were supposed to be learning while they played; get other players in the group to help all of them be better focused; and, as a last resort, suggest that the friend find another partner pair that might help him or her to better concentrate the next time they play.

Dear Etiquette Expert, One of the students in my class always wants to be the one to say what we should do when it is his team's turn. He doesn't give his partner a chance, no matter who it is. We feel like we understand the math, but are just not as quick as he is. How can we help him to understand that his partner needs to be involved in deciding what to do as well?	Dear Etiquette Expert, I am a very shy person and get nervous when we have to find our own partners. Sometimes I go sharpen a pencil or something and feel silly that I don't know how to find a partner myself. Often I wait until my teacher notices that I don't have a partner and she assigns me to someone. Can you help me figure out how I can do this more on my own?
Dear Etiquette Expert, My partner and I were working on a puzzle, but after working on it for a little while she just gave up. She said it was too hard and we should do something else instead. I thought we could solve it if we worked together. How do I convince her to work longer and that the effort is worth it?	Dear Etiquette Expert, My team won the game we were playing today, but my partner made a big deal about our win and I felt really embarrassed. He kept telling the other team that we were much better than they were and that we would win the next time that we played, too. How can I avoid this big scene again?
Dear Etiquette Expert, Sometimes I really like solving problems alone, but my teacher likes team members to work together. I get too distracted when my partner starts talking right away and can't think about the math. I want to ask if we can start a turn by reading and thinking to ourselves, but I don't want to seem too different from everyone else. What do you think?	Dear Etiquette Expert, My partner and I had a lot of questions about the puzzle today. We read the directions again, but it didn't help and everyone else looked like they knew what they were doing. We finally just filled in numbers randomly and said we were done. I know we need to find a way to work through being stuck. Can you help us figure out how to keep going?

Figure 2.4 Etiquette Expert cards

This exploration gave students the opportunity to think about what they could say or do when something wasn't going well for them as they were playing games or solving puzzles. The teacher also found that when these situations arose, asking a simple question, such as *What would the Etiquette Expert suggest here?*, prompted students to be more proactive—to speak up for another player or for themselves.

Let Them Be the Mathematicians

As teachers, we all feel responsible for helping our students learn, but when we don't share that responsibility with them, our own behaviors can interfere unknowingly. Sometimes we might get excited when we notice a particularly good possible move in a game and point it out before the student notices it. Occasionally, we might draw solvers' attention, a bit too quickly, to how two clues could be combined to yield important information about a puzzle. During discussions, we might answer questions to which other students could respond instead. All of these behaviors have the unintended consequences of limiting both our students' sense of responsibility for their own learning, and their development of the habits of mathematicians. We might want to ask ourselves:

- Do I do too much telling, rather than letting my students think about the situation first?
- Do I believe struggle can be productive and let my students struggle long enough before I get involved?
- Do I have shared goals and expectations with each of my students?

No matter how long we have been teaching, we all need to ask ourselves these questions. As mentioned earlier, we encouraged students to be critics and cocreators during this writing. It wasn't long before we asked ourselves why we hadn't previously engaged students in such thinking. As we realized how much mathematical thinking we were doing as we created or adapted games, puzzles, and exit cards, we were even more amazed by this omission. Being in the role of game-, puzzle-, or exit card–creator provided students with opportunities to analyze key ideas as well as build on and critically analyze the thinking of others, which are important mathematical habits of mind. So we learned, once again, that we must make sure we are offering students every opportunity to take responsibility for their own learning. We continue to refer to this idea in each of the following sections of the chapter.

Rules Students Can Decide for Themselves

You may have had the experience of playing a particular game with new friends or with relatives who live in a different part of the country, and discovering that you play the game by slightly different rules. A game of checkers may begin with players establishing whether or not you have to jump, if you can jump. It should be made clear that it only matters that the players agree on the rules; either way, the game is still checkers. One teacher, who has shared this example with his students, replies, "Oh, that's a do-I-have-to-jump question that you decide for yourselves," when players ask questions such as the following:

- How do we decide who goes first?
- Does a team that makes an error lose a turn or just get corrected?
- Does a roll of the dice count if it rolls off the desk?
- Should both teams have the same number of turns, or does the game stop as soon as a team wins?

You'll find that the directions for the games in this book that involve taking turns assume that players will decide who plays or takes a particular role first. We found that students usually relied on rolling a die if one was already needed for the game, or on rock–paper–scissors when a die was not needed for play. When we asked students how they might decide, these were the two techniques they usually shared first. Students also frequently suggested using comparisons about themselves, such as closest birthday, oldest or youngest, name with the most letters, or first names in alphabetical order. You may want to have a conversation with your students and have them make a list of ways they could decide who will go first.

Assessing Learning and Setting Goals

Worthwhile assessment requires teachers to have a clear understanding of what is to be learned, the developmental progression in which the learning is likely to occur, and the evidence that will suggest such learning has been accomplished. It is upon such a foundation that we make decisions as to how particular games and puzzles can accomplish established goals. Creating such goals and making them clear to all stakeholders is essential.

Setting Goals

Teachers, coaches, and administrators spend a great deal of time reviewing data and establishing learning goals. Such goal setting might be at a district, school, classroom, or student level. Shared ownership of such goals is necessary for their success, yet students are not always included. Rich Newman (2012) contrasts his son's ability to articulate goals related to his favorite video game with students' understanding of their learning goals in school. His son could identify goals for his playing and what he must accomplish to meet those goals. He could name recent improvements he had made and how they came about, as well as the next steps he would take to continue growth. He knew his current level of achievement, what he was best at, and specific aspects of the game he found challenging. How many of our students could give such detailed insight into their learning at school?

Some classrooms do have established practices for creating goals and involve students in that process. Consider the following classroom example. Within the structure for setting goals in this classroom, students understand that the games and puzzles they are assigned to or choose to play would be aligned to their learning goals.

The students have completed a practice unit test. The teacher has given them a list of the correct responses and a rubric for evaluating the open-ended response question. In partners, students are to apply the rubric and then reflect privately on two questions: What do you think you know well and why do you think so? *and* What do you need to work on further and why do you think so? *Yan is reviewing his thoughts with his teacher, who concurs with Yan's report that he is progressing well with multiplication and division of positive and negative numbers, but that he needs to continue his work with addition and subtraction. They both notice that he seems to have a good deal of difficulty when he is subtracting a negative number. Together, they set two goals for Yan: to rely on the use of the number line image when he needs to solve such problems until he feels comfortable solving without it, and to be willing to check his answer once he gets one to see if it makes sense. Together, Yan and his teacher decide that two weeks is a reasonable time limit for meeting these goals and that they will meet at the end of week one to check on progress. Yan records these goals in his math notebook, beginning with the phrase* In the next two weeks I will. *He understands that these goals provide an important focus for his learning and that he will be held accountable for meeting them.*

Examining the Evidence of Learning

Once learning goals are established, we must collect evidence as to whether or not students are meeting them. There are a variety of assessment artifacts connected to games and puzzles, but they are not always observed, recorded, or analyzed, and we all know that teachers can't be everywhere or consider every piece of student work. However, we believe there are ways to make such evidence more visible and thus more readily available to inform our instructional decision making. The process begins with the initial introduction of the game or puzzle.

Observations

As soon as students begin to solve or play, we encourage you to become an active observer. In your first observation you want to make sure everyone understands expectations and you want to look for trends or patterns among the students. We suggest specific observations within the "What to Look For" section for each game or puzzle, but, in general, there are four major goals for this initial observation:

- Make sure students are following directions correctly, and intercede as necessary, before misinterpretations are practiced and become more difficult to change.
- Note what students are talking about and look for ways they are relying on one another to reach success. What vocabulary do they use? What questions do they ask one another? Do they offer one another encouragement?

- Look for examples you want to share during the large-group debriefing regarding partial understandings, strategies, or interactions. Ask students' permission to share the examples and note the order in which you want them to do so.
- Look for any challenging situations in which the students are having difficulty getting along while playing a game or solving a puzzle. Sometimes you may wish to address such a situation immediately, while at other times you may prefer to wait.

Consider this teacher's reflection on what she learned by observing her students.

TEACHER REFLECTION

I introduced a new game today and, as usual, circulated as students began to play. Most of the teams jumped right in, excited to begin playing. However, when I walked near one of the groups, I overheard Sammie telling her group that she was frustrated that she didn't understand the directions to the game and that she wasn't sure she could do the math anyway. She told her partner to just go ahead and play without her and she would watch. It was heartening to hear her partner and the students on the other team tell Sammie that they would help her to figure out how to play. They all started the game together and stopped in the middle of each play to be sure that Sammie understood what they were doing. When I checked back later, Sammie was fully engaged in the game and challenging her partner as to their next best move.

This kind of peer assistance was not always the case in my classroom. Students used to play games with one student playing against another, and if one didn't understand how to play then the other just won easily. Now that my students play games as partners, they have realized that it is much more enjoyable to talk about their next moves and to figure out the math together. I was also proud of myself that I didn't get involved in Sammie's dilemma. I used to sit down with Sammie as soon as I saw that she was having trouble and try to help her. Now, it's a pleasure to see my students communicating with each other about what they are learning and I can focus on listening and gathering ideas for follow-up discussions.

As opportunities to play and solve continue, perhaps with variations included, you can focus on one or two groups per session to observe more carefully, taking notes regarding evidence of learning and examples of challenges for individual students.

Student Work

We suggest utilizing two sources of written work: recording sheets and responses to exit cards. Teachers found they could often skim these pieces, looking for correctness or patterns among the group. Sometimes they focused on a particular student's learning, so his or her responses received more attention. For example, Figure 2.5 shows a student response to the second exit card suggested for *Four of a Kind* (page 47), a game in which players match equivalent visual and symbolic representations of ratios. The question asks, "How would you help someone understand that 15 for every 4 is not equivalent to 20 for every 9?" The student's response, shown in Figure 2.5, demonstrates that she understands that ratios can be written and evaluated as fractions. A follow-up question might involve asking her if she can provide another explanation involving ratios.

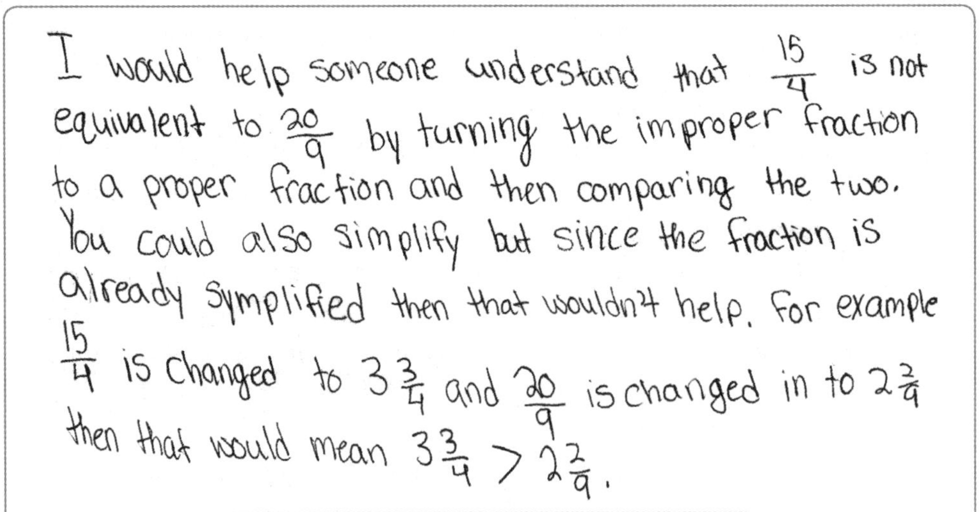

I would help someone understand that $\frac{15}{4}$ is not equivalent to $\frac{20}{9}$ by turning the improper fraction to a proper fraction and then comparing the two. You could also simplify but since the fraction is already symplified then that wouldn't help. For example $\frac{15}{4}$ is changed to $3\frac{3}{4}$ and $\frac{20}{9}$ is changed in to $2\frac{2}{9}$ then that would mean $3\frac{3}{4} > 2\frac{2}{9}$.

Figure 2.5 Student's response to the exit card

At the end of the day, a teacher may have a pile of recording sheets and wonder what to do with them. Students often do not receive feedback about such work and often regard recording sheets as just a place to "do their math" while they are playing the game. One teacher told us, "If a recording sheet is provided, I always have my students use it. They know that it is important for them to remember what they were thinking while they were playing. They also know that I will review the sheets and respond frequently." We agree. When teachers provide feedback, students value the recording sheet more; they view it as a tool for communicating their thinking and demonstrating what they know.

There are a variety of ways you can use the exit card questions suggested in the game chapters:

- Use the questions over time, choosing one question the first time students investigate a game or puzzle, and another after students are more familiar with the game or puzzle.

- Choose one question to give to the whole class, or vary the questions for different students based on their readiness or choices.
- Provide more than one question and let students choose which to answer.
- Include one of the questions on an at-home assignment or formal assessment, letting students know that games and puzzles are an integral part of their learning and require accountability.
- Offer one question twice, once after the first exposure to the game or puzzle and then again after further explorations, to document any changes. You may want to have students compare their two responses and reflect upon how much they have learned.

After collecting responses to both exit cards and recording sheets, you could do one of the following:

- Choose examples to share with students during the next math lesson.
- Sort responses quickly into three piles—not proficient, working toward proficiency, and proficient—and use the results to differentiate instruction.
- Share a simple, generic rubric (see, for example, Figure 2.6) to help clarify expectations, after which you or your students could apply it occasionally to their responses.
- Create a task-specific rubric with a colleague at your grade level and work together to apply it to students' responses.
- After reviewing them with students, place two responses to the same question or task, completed over time, in the students' portfolios.

Concepts	There is no evidence of conceptual thinking, or thinking is incorrect.	Partial understanding of concepts is demonstrated.	Full understanding of concepts is demonstrated.
Skills	There are several computational errors.	Most computation is correct.	All computation is correct.
Communication	There is no use, or mostly incorrect application, of mathematical terms and symbols.	Mathematical symbols are used correctly and some relevant vocabulary is included and used correctly.	All relevant mathematical symbols and vocabulary are included and used correctly.
Communication	No, or only incorrect, examples or explanations are provided.	An incomplete explanation is provided, or more clarity is needed.	Explanation is clear, correct, and includes examples.

Figure 2.6 Generic rubric for exit questions and recording sheets

Fostering Productive Discussions

Math talk is now a common term, illustrating the expectation for communication in today's math classrooms. *Accountable talk* describes the specific kind of discourse in which we want our students to participate. Such talk requires students to ask each other about their thinking, listen to each other carefully, build on the ideas of others, and use evidence to justify their ideas. It also means creating classrooms where students are willing to share their initial thinking, not just their finished products. We want to build classroom communities where errors are viewed as part of the learning process, because using errors for instructional purposes has the potential to increase students' understanding (Bray 2013). Such instructional opportunities are dependent on finding tasks that allow us to tap into potential misunderstandings and creating classrooms where errors are valued as learning experiences rather than something to be avoided because of the negative connotation of making them.

To some extent, such conversations, along with the teacher talk moves suggested by Chapin, O'Connor, and Anderson (2009) and Kazemi and Hintz (2014), are incorporated into many classrooms. Our task here is to think about connecting this talk to games and puzzles.

First and foremost, such conversations depend on significant tasks. Engaging games and puzzles provide something for students to talk about. Whole-group debriefing sessions offer opportunities for students to describe the mathematical understanding they gained from the game or puzzle, discuss strategies, and explain solutions. We want students to hold these same types of conversations while they are playing and solving in small groups. Sentence frames can be used to help students participate. Examples are shown in Figure 2.7.

Ask questions when you don't understand.	Can you explain why this move ...? What do you mean when you say that ...? Can you help me to understand ...?
Ask about the thinking of others.	What do you think we should ...? How did you decide ...? Do you want to say anything about ...?
Make predictions.	I predict that ... If we do... then I think ...
Build on the thinking of others.	I agree with ..., but ... I use this idea, too, when I play, but I ... _______'s idea makes me think ...
Give and ask for evidence of thinking.	An example from the game we played is ... I think ... because ... Can you tell me why it is true that ...?
Look for patterns and generalizations.	Now I am wondering ... This puzzle reminds me of ... because ... We could also use this when playing ...

Figure 2.7 Sentence frames for games and puzzles

Meeting Individual Differences

Your assessment data will help you identify mathematical readiness levels. Several specific ways to meet such differences are suggested in the game chapters within the "Variations" sections. Ideas for meeting other differences are provided in the "Tips from the Classroom" sections. Here we would like to provide a general list of ideas.

- With permission, create a video of students playing the game that other students may watch as a way to see the directions in action. This can be particularly helpful for students who learn best by example, rather than by oral or written directions.
- Create sound barriers for rolling dice by having students use rug samples or a piece of felt on which to roll.
- Cards and dice come in a variety of sizes, colors, and textures. Offering a variety allows students to choose what's best for them.
- Some students may not have time to complete an exit card during class. You could ask them to, or students could decide to, complete or refine their responses as part of their homework.
- As students play and solve together, some may prefer to respond to exit cards together as well. You can vary expectations for cooperative and individual responses.
- Students react in different ways to competitive play. Though it can be motivating for many, some become overly competitive and others resist competition or become quite anxious. Nearly all games can be played cooperatively, with the goal of trying to improve by working together.
- Some students may not be familiar with how to shuffle cards. Show them how cards can be cut and rearranged several times or spread out on a flat surface and moved around before re-forming a deck.

Organizing for Success

After many years of thinking that the best way to introduce a new game or puzzle format was to demonstrate it for a few students and then let them introduce it to others, we now believe that a new game or puzzle format should be introduced by the teacher to the whole class or a large instructional group. You'll find that each of the games and puzzles in this book is introduced in this manner. This does not mean that you always have to play an entire game or completely solve a puzzle in a large-group setting. Sometimes you can consider a miniature version of the game or puzzle, take just a few turns or look at a few puzzle clues as a class, or explore or review related mathematical ideas and relevant terms. We believe such introductory experiences help students in the following ways:

- They support the notion that students are part of a diverse learning community.
- They model and reinforce expectations for how best to play games or solve puzzles with team members—that is, to be patient with opponents, persevere, and be gracious winners and losers.
- They model and reinforce ways to participate in accountable talk related to games and puzzles.
- They increase the likelihood that students will be able to follow directions and meet expectations during opportunities to play or solve.
- They provide you with occasions to gain the formative assessment data you need to differentiate follow-up opportunities for students to play the game or solve similar puzzles.

Grouping Students

When games and puzzles are just for practice, teachers often group students homogeneously and have one student compete against another or have students complete a puzzle alone. When the goal is to build a deeper understanding of mathematics and mathematical habits of mind or practices, heterogeneous grouping makes more sense. Playing games and doing puzzles can provide wonderful opportunities for students with different abilities to discuss a common goal. Students may be surprised to find that classmates who they thought were struggling with a concept have interesting strategies to offer or connect clues that others hadn't thought to try. Thinking of all students in the room as mathematicians, rather than just those students who seem to be the fastest and most verbal about what they know, brings new light to learning opportunities.

As with all classroom activities, you always have the choice to assign students to teams, pick opposing teams, and select locations for play, or you may choose to have the students make such decisions independently. Though we support having students make these decisions whenever possible, we often lean toward having the teacher make partner decisions when the class is first considering a new game or puzzle format, assigning students in ways that will best support their learning. You can display the information for all to see so that little time is lost in finding co-players. We recommend that students play or solve, in teams, immediately after they have experienced a whole-class introduction.

Having Students Play and Solve in Teams

When the goal of playing a game or solving a puzzle is to deepen learning, rather than only to practice skills, we recommend that students play and solve in teams. All game directions assume that two teams are playing, composed of two players each. While we recognize that, for a few students, this may be too challenging, we consider it optimal in nearly all cases. We also recommend that puzzles be solved in a group. We don't define group size, as you know what is best for your students, but two or three students working together is usually ideal. Sometimes, particularly with puzzles, students need a few minutes to work independently before teaming, and this is fine. We've found that some students prefer to read and try to interpret a clue or two by themselves but then, without prompting, begin to work with their partners. Some teachers build in independent think time before cooperative thinking, and some groups build this in for themselves. We remember Jason, who suggested to his group, "Let's read to ourselves first and then we can talk."

Some teachers organize teams before introducing a game or puzzle; others draft team combinations, perhaps by using magnetized name cards on a whiteboard, that can be quickly altered based on assessment data gathered during the introductory experience. Other teachers prefer to quickly organize teams with oral directions, and some allow students to organize themselves. Regardless of how teams were organized, we have been privy to wonderfully unique and informative mathematical conversations by listening to students play and solve in teams. Consider the following reflection from one of the teachers who worked with us.

TEACHER REFLECTION

Traditionally the kids in my class would play games against another player. They didn't want to talk about the math because they didn't want to give away their strategies, especially when there was strong competition. I was surprised by the impact of having students play in team pairs. Immediately the noise level in my room increased in a good way. The students needed to talk about their strategies with their partners and they would debate about which strategy was best. I also found that kids who didn't have strategies benefited from their partners' think-alouds, and over time, they began to try out the same strategies on their own.

Regaining Students' Attention

The engagement with a good game or puzzle can lead to a higher level of noise in the classroom than some teachers would prefer. No doubt you have one or more signals that remind students to use their classroom voices or that alert them to immediately stop and pay attention to you. When you introduce such a system, likely during the first week of school, it is important that students practice the expected response while playing a game or engaging in solving a puzzle. Interrupting such activities may be challenging for them, especially if they feel as if they are about to win or reach a collaborative goal or solution. Assuring students that they can return to the game or puzzle after you have finished talking or at another time will often facilitate this process.

Making Directions Accessible

Frequent requests to repeat or provide another copy of directions can be frustrating. As you know, this is not a good use of your time, nor does it teach students to listen, remember, or organize their own materials. To avoid these distractions, you could:

- Establish an area in the room where directions are available; a reproducible with directions is provided for each game or puzzle in this book.
- Make a video of you or your students (with permission) reading the directions, and make it available on the classroom website or for viewing at a designated area in the classroom.
- Encourage students to write a short, personalized version of the directions they can keep with them while they are playing the game for the first time.

Managing Materials

Puzzles often do not require materials other than the puzzles themselves, but games usually do. A designated area of well-organized materials makes it more likely that students will respect and return game components appropriately. Having students play or solve in groups simultaneously requires multiple copies of materials and containers.

If your materials are well cared for and organized, it sends a message that playing games is important and that you trust students enough to let them use materials that are attractive. Here are some questions and responses that can guide your thinking about the organization of game materials in your classroom.

- *Does the organization and labeling encourage student independence?* Clear labels on containers that are organized in a systematic way allow students to find and return materials more easily. Game directions along with a list of required materials pasted inside each game container help ensure that components won't get lost and that students will meet expectations for play and for cleaning up.
- *Does the organization support long-term game use?* Placing number cards in a soap dish within a box protects the cards, as does laminating any paper materials.
- *Does the organization support large-group use?* Large-group participation requires multiple copies of materials. Keep multiple sets together in easily carried containers when preparing to introduce a new game. Some of these copies can then be moved to a game library for continued use at another time.
- *Does the organization maximize instructional time?* Having extra sets of frequently needed materials, such as dice or cards, allows students to replace missing components immediately, when needed.

Working with Families

We know that the amount of parent involvement correlates to student achievement; this also applies to games and puzzles (Kliman 2006). Some families may be surprised that their early adolescent children are playing games in class. You can communicate why the use of games and puzzles is important in the classroom through an e-mail message or a letter, perhaps including the directions for a popular game or the questions that students must answer on an exit card after completing a puzzle. A sample of an introduction to such a message is provided in Figure 2.8.

Dear parent(s) or guardian(s),

Your children will be working hard this year to understand the mathematics they are learning, compute efficiently, and be strong problem solvers. Oftentimes we will use games and puzzles to support this learning. Students will play and solve in groups so they can talk about their thinking and discuss strategy. I use exit cards (short questions students complete after playing a game or solving a puzzle) and recording sheets to make sure the students are recording what they are learning while they are playing or solving. I often provide feedback on the recording sheets and exit cards so that students can learn from what they are writing. You might find that your children are assigned to play a game at home in order to answer a homework question or to prepare for a quiz or test, and they may ask you or another family member to play with them. As a result of this activity your children will learn to talk about the math they are learning more articulately, compute more efficiently, gain number sense, and become stronger mathematicians.

To give you an initial idea of the types of games and puzzles we will be using, I am sending along ...

Figure 2.8 **Sample note to parents and guardians**

Involving families in game playing or solving a puzzle gives both students and family members the opportunity to share the math together in a way that is different from completing a worksheet or doing problems from a textbook. Oftentimes parents are not aware of the academic language we are using in schools to talk about mathematics, and highlighting vocabulary for parents will positively affect how they communicate with their children about mathematics.

Playing a game with a sibling, older or younger, or with a parent allows the student to take a leadership role in explaining the rules and sharing what he or she has learned about the game or puzzle and the mathematics involved. The conversation about the math the students are learning is embedded in a meaningful context, and parents are likely to gain more information than they get in response to the question *What did you do in school today?*

You can establish a lending library in your classroom, allowing students to bring games home. One method for providing structure to this responsibility is to include a materials list with each game packet that must be checked off when the game is returned. Students can check out the game from the classroom games library just as they would a library book, writing their name on an index card and placing it in a file or container designated for this purpose. Include a comment card in the bag so that students can write something about an interesting conversation they had, something new they learned, or questions they had while they were playing. This card could be shared with the next students who explore the game.

You can also require students to play a game or solve a puzzle as part of their homework, or as practice toward an established goal. Students may also be encouraged to take home a particular game in order to prepare for a quiz or a test. You might choose to require students to record the work of a family member while playing the game or to describe a challenge a sibling encountered when solving a puzzle. Encouraging written communication about what transpired at home will provide students with an opportunity to solidify their thinking about the mathematics content.

Conclusion

We know you recognize that the instructional practices addressed in this chapter relate to all aspects of your teaching, but we hope that you can now envision them more distinctly in game- and puzzle-related situations. We also hope that you'll discover new joy in using these games and puzzles or some of your old favorites with these ideas in mind. Your students will appreciate the opportunities to learn in this manner and you will see growth in their mathematical understanding, computational fluency, and algebraic thinking.

The Number System

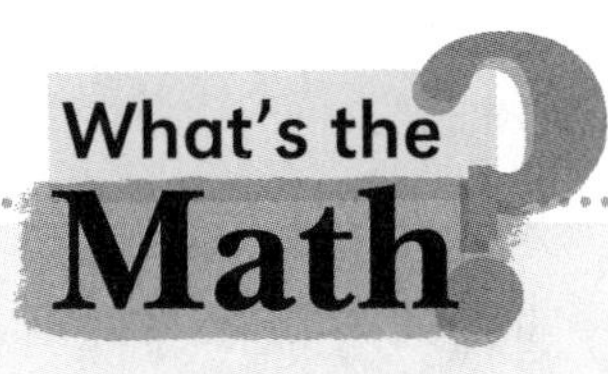

At the middle school level, students solidify, extend, and apply their understanding of our number system. While full proficiency is expected, including the use of formal, algorithmic procedures, it takes time for students to develop such skills, which must be grounded in conceptual understanding (Choppin, Clancy, and Koch 2012). Knowledge of why the algorithms work supports students as they transition from operating with whole numbers, fractions, and decimals to the inclusion of variables.

Students in sixth grade are expected to fluently add, subtract, multiply, and divide whole and decimal numbers, and their work with fractions is extended to the computation and interpretation of quotients. With decimal numbers, computational approaches are quite similar to those with whole numbers. However, operations with fractions, and division of fractions in particular, prove to be more challenging (Park, Güçler, and McCrory 2013). Students often struggle with these ideas, even though they have previously worked with whole number division and other operations with fractions. Sixth graders also identify greatest common factors and least common multiples, and work with rational numbers as points on number lines.

In seventh grade, students extend their exploration of, and must reach fluency with, addition, subtraction, multiplication, and division of rational numbers. Students develop computational strategies based on the properties of these operations established in the elementary years.

Over seventh and eighth grades, students convert between rational and decimal number representations and in eighth grade, students are introduced to irrational numbers.

In its final report, the National Mathematics Advisory Panel indicated that, of all the skills needed for success with algebra, it is proficiency with fractions, decimals, and percent that is lacking (NMAP 2008). Games and puzzles in this chapter provide students with opportunities to recognize equivalent representations of numbers, to compare numbers presented in a variety of formats, to develop computational fluency with integers and rational numbers, and to strengthen the precision of their mathematics vocabulary related to number and operations.

Why This Game or Puzzle?

It is difficult for middle school students to make connections among the many forms and representations of rational numbers (Beyraneyand 2014). For this puzzle, students are given sixteen individual puzzle pieces that must be matched on all sides. Different representations of an equivalent fraction, decimal, and percent make a match. To reach the solution, solvers must decide how to sort and organize the information—decisions that are key in real-world situations.

***Match It* Puzzle A Pieces**

250% 0.5%	2.5 0.6 0.3	60% $\frac{3}{100}$ 0.03	0.03 7.5%
0.005 1.5 75%	$\frac{3}{10}$ 150% 3.4 50%	3% 340% 0.7 $\frac{1}{5}$	0.075 70% 25%
$\frac{3}{4}$ 0.02 0.4	$\frac{1}{2}$ 2% $\frac{1}{10}$ 4%	0.2 0.1 $\frac{2}{5}$ 5%	0.25 40% 7.5
$\frac{2}{5}$ 20%	$\frac{1}{25}$ $\frac{1}{5}$ 0.01	$\frac{1}{20}$ $\frac{1}{100}$ 2%	750% $\frac{1}{50}$

Three versions of the puzzle are provided. The first puzzle requires students to match an equivalent fraction to a decimal or a percent representation. The second version varies the puzzle by including "greater than" or "less than" comparisons; and the third puzzle offers students the opportunity to relate scientific notation and exponent forms of a number.

Math Focus

- Identifying fraction, decimal, percent, scientific notation, and exponent form equivalence

- Comparing fraction, decimal, percent, scientific notation, and exponent forms of a number

Materials Needed

- 1 *Match It* Puzzle (A, B or C) per pair (or group) of students (pages A-4 to A-6)
- Optional: 1 *Match It* Directions per pair (or group) (page A-8)

Directions

Goal: Arrange each puzzle piece so that the numbers represented on all adjoining sides match.

- Work together.
- Organize each of the 16 puzzle pieces into a square.
- The pieces should be placed so that the numbers on each of their sides match; that is, the representations on the sides of any adjacent squares have the same value.
- Check to make sure you have matched each side correctly.

How It Looks in the Classroom

One sixth-grade teacher puts the decimal *0.15* on the board and asks the students to write, on a sticky note, another representation for this number. As students are writing, the teacher circulates to choose some of the representations to put on the board. As the notes are placed, students are asked to notice which representations seem similar, and which seem different from others.

One student writes the words *fifteen hundredths*, while another shows the equivalent fraction $\frac{15}{100}$. Maggie draws a picture of a sign showing "15% off" at a sneaker store, and David suggests, on his sticky note, *You could write any fraction that is equal to* $\frac{15}{100}$. The class discusses how the representations on the sticky notes might be categorized and the teacher is pleased to hear the students use appropriate vocabulary as they equate hundredths to thousandths and discuss that the percent symbol represents a fraction with a denominator of 100. The teacher compliments the students on their varied responses and then draws their attention to the board, where she has displayed just four cards similar to those in the puzzle (Figure 3.1).

Card 1	Card 2	Card 3	Card 4
(right) $\frac{5}{4}$ (bottom) 0.125	(top) 300% (left) $\frac{2}{16}$	(left) 1.25 (bottom) 3	(top) $\frac{1}{8}$ (right) 12.5%

Figure 3.1 Mini-*Match It* puzzle

The teacher asks Gauri to choose one side of one card, and she chooses the side with $\frac{1}{8}$. She asks Gauri to call on another student to find a different representation for $\frac{1}{8}$, and Eli chooses 0.125. The teacher asks the class if the two puzzle pieces may be put together, with the end goal of matching all of the pieces and making a square. Maggie agrees that the two representations match; however, she also thinks that $\frac{2}{16}$ could be matched with $\frac{1}{8}$, and others concur. The students are intrigued by the fact that there are two representations that are equivalent to $\frac{1}{8}$, though there is only one way to make the puzzle pieces fit together.

The teacher moves the pieces showing $\frac{1}{8}$ and 0.125 together, placing the matching sides next to each other. Students continue to volunteer to complete the four-piece square, following along with matching sides with equal numbers but different representations. She asks the students to describe the completed puzzle as shown in Figure 3.2 and they talk about how there must be matches on all sides. She then tells the class that each group will be given a set of cards similar to those displayed, but with more options, and students must match all the cards to form a square. The students are eager to begin.

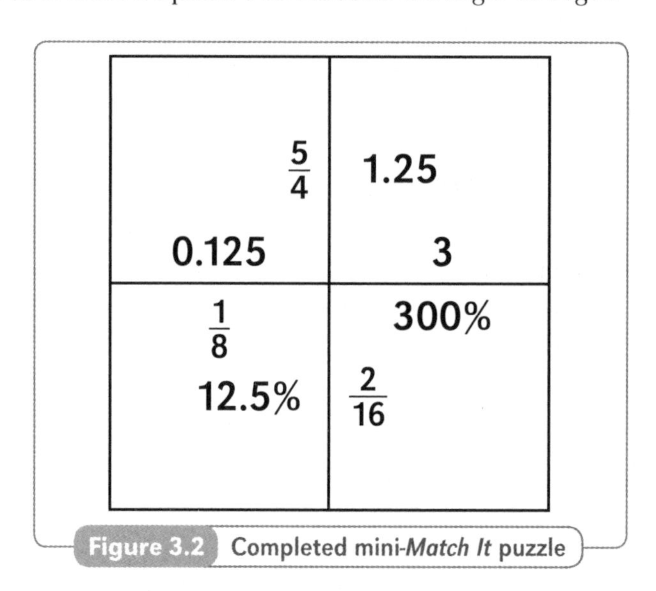

Figure 3.2 Completed mini-*Match It* puzzle

Tips from the Classroom

- Getting started is sometimes a challenge for students. Promote student conversation about where to begin and whether they want to split up the pieces to find matches or work together to match each piece. Try not to intercede in the conversation as they work through their strategies for solving the puzzle.
- Encourage students who may need to visualize a number represented differently from the way it is presented to use a manipulative or draw their own models on paper.
- Some puzzle sides have more than one match. Students need to choose the piece that allows all sides to match. Some students did not notice this and thought there was an error in the puzzle when they were left with unmatched pieces. You may need to assure them that the puzzle works; or, in some cases, encourage them to continue solving by asking them to see if there are pieces that could be exchanged.

What to Look For

- Do some students have difficulty finding a particular type of representation to match another, such as a percent? If so, you may want to offer a small-group instructional opportunity at another time.
- Do students confuse terms that are similar except for their place value, such as 0.2 and 0.02?
- Do students work together as a team to solve the puzzle, or do they try to work individually to match pieces?

Variations

- You can create less and more challenging examples of this puzzle by restricting or expanding the representations.
- You can make the puzzle more challenging by putting numbers and representations on all sides of the puzzle pieces, requiring students to identify the pieces that have sides without a match as the corners and edges.

Exit Card Choices

- You are creating a fraction and decimal jigsaw puzzle. You want the solver to have to match 0.24 with a fraction. Write several representations in fraction form that would make a match.
- A friend has matched two puzzle pieces on one side but can't find two puzzle pieces to put above these pieces. He says, "This puzzle doesn't work." What might you say to your friend to be helpful?

Extension

Students can create their own versions of this type of puzzle using the Make Your Own *Match It* Puzzle Template (page A-7).

Why This Game or Puzzle?

Estimation skills are essential; we probably use them every day. We estimate when exact answers are not required and to check the reasonableness of computation completed with paper and pencil or an electronic device. For instance, in a grocery store, we may estimate to see if we have enough money for what we are purchasing and then estimate again to make sure the total the cashier tells us makes sense.

Tic-Tic-Tac-Toe Game Board A

Dividend		
315		7,875
	18,900	
2,835		1,260

Divisor		
9		35
	105	
15		3

3	140	6,300	81	525
84	180	315	2,625	9
540	189	75	35	420
945	225	21	12	2,100
875	105	36	1,260	27

According to Lan and colleagues, "Computational estimation is a complex skill involving many of the same subtleties and complexities as problem solving," (2010, 1). Yet typically, most of the opportunities students have to practice their computational skills are focused on finding exact answers.

In this game, students are given two sets of numbers. They choose one number from the set of dividends and one from the set of divisors, divide, and write an *X* (or an *O*) on the quotient on the game board. The goal is to write an *X* (or an *O*) on four adjacent quotients in a row, column, or diagonal. This aspect of the game encourages students to estimate quotients when choosing dividends and divisors, make conjectures, plan ahead, and decide when they want to block an opponent's move. When we field-tested it, students told us to make the game board look more like a tic-tac-toe game, so we eliminated the outside borders. Three versions of the game are included; each is focused on a different type of number. Game Board A offers whole numbers, Game Board B provides fractions, and the focus of Game Board C is decimals.

Math Focus

- Dividing whole numbers, fractions, and decimals
- Estimating quotients

Materials Needed

- 1 *Tic-Tic-Tac-Toe* A, B, or C Game Board per group (pages A-9 to A-11)
- Optional: 1 *Tic-Tic-Tac-Toe* Directions per group (page A-12)

Directions

Goal: Choose divisors and dividends in order to mark off four quotients in a row, column, or diagonal on the game board.

- Decide which team will be *X* and which will be *O*. The first team picks one number from the set of dividends and one from the set of divisors. Both teams compute the quotient. (Teams get to pick only once, even if they discover that they don't get a quotient they want.)
- Once both teams have confirmed the quotient, the first team finds it on the game board and writes the team's mark (*X* or *O*) in that space.

- If the team gets a quotient that is already marked with an *X* or an *O*, it loses its turn.
- Teams alternate turns.
- The first team to write its mark in four touching quotients in a row, column, or diagonal is the winner.

How It Looks in the Classroom

One sixth-grade teacher displays the expression $7\frac{1}{5} \div \frac{3}{10}$ as well as the three numbers 24, $\frac{1}{5}$, and $\frac{24}{5}$. She says, "I am going to count slowly to three. By the time I get to three, I want you to have chosen which of these three numbers you think is the quotient." As she counts, she notes the expressions on the students' faces. She notices that some look confident as they are thinking, while others look somewhat confused.

Joshua reports that he thought it was $\frac{1}{5}$. He explains, "It is the only number less than $7\frac{1}{5}$ and you get smaller numbers in division." Shaylyn tells how she eliminated $\frac{24}{5}$ right away because the fraction is about five and she thinks that there are a lot more than five three-tenths in seven. Diego says, "It has to be 24, because that's the greatest number." Many students agree with Shaylyn and Diego, though a few of them agree with Joshua, and some share that they were confused because they were sure the number had to be a fraction. The teacher asks them to work with their neighbors to identify the correct quotient and explain why it makes sense. After working in pairs, the students discuss their findings and Leighella reports, "I think 24 makes sense because the problem is asking how many three-tenths there are in $7\frac{1}{5}$ and if there are three three-tenths in one, then there are at least twenty-one three-tenths in seven."

The teacher displays the *Tic-Tic-Tac-Toe* Directions (page A-12), tells the students that predicting quotients is a focus of this game, reviews the rules of play, and makes sure students understand what it means for quotients to be "touching." Players then form teams, get copies of the game board, and begin playing.

Tips from the Classroom

- Some students are likely to begin by choosing numbers randomly. It's best not to intervene too quickly. After several turns, players often use estimation to make better choices for divisors and dividends.
- It can be helpful to ask players to think aloud as they choose their numbers, as their thinking gives other students access to a variety of strategies.
- You may wish to allow some students to use calculators to check quotients.
- Some students may decide to find all of the quotients before playing the game. Help them understand that the purpose of the game is to choose quotients within a short time span, encouraging estimation instead of computing exact answers before making choices.

- When we were field-testing this game, some students forgot that they needed to choose one number from each set and instead, chose two numbers within the same set; for example, they chose two dividends instead of a dividend and a divisor. They then thought the board was missing a number. You may need to remind students that they need to choose one number from each set.

What to Look For

- Do students choose randomly or do they use strategies for making choices? For example, do they round, consider friendlier numbers, apply the inverse relationship between multiplication and division, or use information from previous computations to inform new choices?
- What mathematical language do students use to describe their thinking?
- What strategies do students use to help them get four touching numbers in a row, column, or diagonal?
- Do students remember to block opponents when necessary?
- What do teams do if they disagree about a quotient?

Variations

- In our field-testing, one group of students suggested this game rule: If a team calculates a quotient incorrectly it loses its turn. This rule motivated opposing teams to check calculations.
- By changing the numbers in the sets and on the game board, you can vary this game, having students find products or percentages. You could also include positive and negative numbers.

Exit Card Choices

- What strategies did you use to choose your divisors and dividends on your turns? Give a specific example.
- When did you think it was important to block an opponent rather than find a quotient that touched the quotients you already had?
- What did you do if you and your opponents disagreed on a quotient?

After playing the game with fractions, a student provided the response shown in Figure 3.3 to the first exit card question. Note the student's use of the inverse relationship between multiplication and division, the translations between fractions and decimals, and the application of what she knows about money.

I needed 1 3/4 to block our opponent. So I changed 1 3/4 to the decimal 1.75. Then I multiplied it by the divisor numbers and hoped to find a number in the dividend box. I Knew that 2x1.75 would equal 3.50 because \$1.75 + \$1.75 = \$3.50. I also Knew that 3.50 is the same as 3 1/2. I tried to use what I Know about multiplication to solve the division problem.

Figure 3.3 Student response to the first exit card

Extension

Have students create, exchange, and play their own versions of the game. They choose five numbers for each set, compute the related twenty-five possible quotients, replace any combinations that don't provide unique quotients, and then randomly write the quotients on the game board.

Game adapted from *Estimation Bingo* in *Zeroing in on Number and Operations: Key Ideas and Common Misconceptions, Grades 5–6*, by Anne Collins and Linda Dacey. (Portland, ME: Stenhouse, 2010).

Three-For

Why This Game or Puzzle?

It takes time for students to understand operations with positive and negative integers. In particular, students may struggle with interpreting the minus sign and develop the habit of using language such as "negative negative five," rather than the more useful "the opposite of negative five" (Bofferding 2010). Frequent opportunities to discuss expressions involving integers can support students' development of more productive terminology.

In this game, playing cards represent negative (red) and positive (black) target numbers. Teams must then find three numbers on the game board to use in an expression with a value equal to the target number. Additional points are scored if the numbers chosen are adjacent on the board, which encourages students to consider a variety of options. As they do so, they are finding the values of numerous expressions and talking to their teammates about how to read and interpret the expressions they create.

Math Focus

- Computing with positive and negative integers

Materials Needed

- 1 *Three-For* Game Board per team (page A-13)
- 1 standard deck of playing cards with face cards and jokers removed; aces = 1
- 1 *Three-For* Recording Sheet per team (page A-14)
- Optional: 1 *Three-For* Directions per team (page A-15)

***Three-For* Game Board**

-1	7	22	-4	5	-25
1	-18	9	20	18	6
10	-19	-7	8	-12	-16
-14	25	11	-15	2	-8
-10	14	-9	19	-16	3
-21	0	-11	6	-3	-20

Directions

Goal: Create sets of three cards each that equate to a target number.

- Give each team a copy of the game board.
- One team draws a card from the deck and places it for all to see. If the card is red, it is a negative integer; if the card is black, it is a positive integer.
- Teams are to find three numbers on the board that can be used in an equation (all four operations are allowed) to equal the target number.
- When they find the three numbers, they call "Three-For" and the other team must stop looking for a solution. They should record their equation on the recording sheet.
- The equation is checked by the opposing team. If it is correct, the team that found the three numbers gets 1 point and crosses the three numbers off the board. Additional scoring will occur if the team finds three numbers in which two are adjacent, allowing two points to be scored, and three points if all three are adjacent.
- If the equation is incorrect, both teams return to finding a solution.
- Play continues until there are no more numbers to choose from or there are no more possible "*Three-Fors*" when a target number is displayed.

How It Looks in the Classroom

As students enter their seventh-grade classroom, each is handed a number card. The teacher brings the class together briefly to discuss the warm-up for today's lesson by stating, "I have provided each of you with a card. If you have a red card, the number is negative. If you have a black card, the number is positive. I will put a target number on the board. You must try to find students in the room who have cards that you can group together to create expressions with a value equal to the target number. You will have six minutes to cooperatively find as many expressions as you can for the target number. Ready, set, go!"

As the teacher circulates, she enjoys listening to how her students discuss not only the rules for integer operations, but why the rules work. She hears Melissa tell another classmate, "Remember we learned that $3 \times -4 = -12$ because that means three groups of negative four." Kai wants to check his thinking about $4 - (-5)$, so the teacher tells him to get a number line template from their classroom materials. Shortly thereafter she hears him say, "If I were subtracting a positive number I would move five left, so I move five right instead."

When the six minutes are up, the teacher asks the students to share some of their expressions for the target number -12. One group shares their expression $3 \times 4 \times (-1)$, but Aidan is concerned, thinking that the answer is 12. When asked why, he shares, "Well, I thought that you would multiply both the three and the four by negative one, and that would make positive twelve." The teacher is pleased that this misconception has been stated and a conversation ensues about the order in which numbers are multiplied and how these parentheses do not necessitate the need to use the distributive property.

After several other examples, the teacher believes that the students are ready to play *Three-For* at their tables. She explains how the game is played and relates the work involved in the warm-up problem to the game. Students are eager to see what target number they will start with as they begin round 1.

Tips from the Classroom

- When field-testing this game, we noticed that opposing teams were often eager to check the work since, if it were incorrect, there would be more time to try to find a solution. Some students suggested that, if they found an error in an opponent's computation, they would receive the point instead, or the opponent would have to lose a point.
- Some students who were still working on understanding computation with integers wanted to use a manipulative, such as colored chips or a number line, to help them determine accurate answers. We encourage the use of these models while students are playing the game, to support conceptual understanding.

What to Look For

- What strategies do you hear discussed that you want students to share in a whole-group summary discussion following the game?
- Do some students need to use a manipulative or a number line for a longer period of time than others when playing the game?
- How well do students persist in their thinking regarding choices for the target number? Are some students ending their turn too quickly, needing some encouragement to continue looking?

Variations

- Change the numbers on the game board to greater numbers, such as those from 30 to 100, including both positive and negative integers.
- Vary the amount of numbers they may use and create a different scoring system, such as one in which students count the numbers they chose to identify a score for that round.
- Other rules might include requiring students to use at least one positive and two negative numbers, or vice-versa, in their choices.

Exit Card Choices

- You are playing *Three-For* and the target number is -9. You think you'd like to use the 3, but need two other numbers. Name four possibilities of number combinations you could use to make -9.
- There are only three numbers left on the board: -4, 5, and 8. What target number would you like to get in order to win a point? If possible, try to find at least one positive and one negative target number that would work.

The student response shown in Figure 3.4 for the second exit card depicts the student's interest in finding more than one positive and negative target number that works.

$(^-4 + 5) - 8 = ^-7$

$(5 + ^-4) \cdot 8 = 8$

$(8 - 5) + ^-4 = -1$

$5 + (8 \div ^-4) = 3$

Figure 3.4 Student response to the second exit card

Extension

Change the game to a puzzle, with each number on the game board used exactly once. The solvers use the playing cards to identify twelve target numbers and then try to create an expression, using three numbers on the game board, for each target number.

Why This Game or Puzzle?

"Developing mathematics vocabulary knowledge allows adolescents to expand their abstract reasoning ability and move beyond operations to problem solving," (Dunston and Tyminski 2013, 40). Helping students develop a precise mathematics vocabulary is important, as an inadequate academic vocabulary can negatively impact content knowledge. Such attention is particularly important at the middle school level, where there is a significant increase in the types of numbers considered and, therefore, the terms associated with numbers and the structures within our number system.

Communicate It! Cards

INTEGER	RATIONAL	IRRATIONAL	REAL
SQUARE NUMBER	CUBIC NUMBER	SQUARE ROOT	CUBE ROOT
WHOLE NUMBER	NATURAL	POSITIVE	NEGATIVE
SCIENTIFIC NOTATION	ABSOLUTE VALUE	DISTRIBUTIVE PROPERTY	COMMUTATIVE PROPERTY
ASSOCIATIVE PROPERTY	IDENTITY PROPERTY	MULTIPLICATIVE INVERSE	PRIME FACTORIZATION

In this cooperative game, students use words and gestures to allow other players to identify a specific mathematical term. Gestures give students a physical reference for a term and help them remember the definition. This game may be particularly helpful to English language learners.

Math Focus

- Using number system vocabulary

Materials Needed

- 1 deck of *Communicate It!* Cards per group (page A-16)
- Timer (sand, phone, or kitchen timer)
- Optional: 1 *Communicate It!* Directions per group (page A-17)

Directions

Goal: Guess the vocabulary word.

- Each team chooses a card from the deck, silently reads the vocabulary word, and plans privately how they will communicate the term.
- The timer is set for 2 minutes.
- Team 1 uses words or actions to communicate its vocabulary word to the other team, without saying the word itself.

- If Team 2 is able to guess the vocabulary word before the timer runs out, the teams get 1 point.
- The other team then tries to communicate the word on its card.
- Play continues until each team has had the opportunity to try to communicate four vocabulary terms.
- If, together, the teams earn 6 points, they win the game.

How It Looks in the Classroom

Many of the students in this seventh-grade classroom enjoy having whole-class discussions about the math they are learning. So, they are excited when their teacher tells them that they are about to play a game that involves standing in front of the class and using words and gestures to communicate vocabulary words. He asks for a volunteer to choose a card from the vocabulary deck and not show the word on the card to the class. The card stating RECIPROCAL is chosen and the teacher asks the student to use words and gestures to communicate the meaning of the word to his classmates. The key, the teacher tells the class, is that the student cannot use the word in the description. The first student in the class who can guess it will choose the next card.

A number of students volunteer to communicate the chosen word; the teacher calls on a student who does not usually volunteer. Maribel stands at the front, somewhat timidly at first, and then places her hands horizontally, one on top of the other, while saying, "Change the parts." Students call out such vocabulary words as *fraction*, *numerator*, and *denominator*. Maribel shakes her head, but her eyes are lighting up as she attempts another gesture. She puts her hands in fists, one on top of the other, and then exchanges their positions so that the other fist is on top and says, "Flip them." Peter calls out, "Reciprocal!" Maribel, pleased that a classmate has guessed the word, looks delighted. The teacher asks Peter to describe how he knew that Maribel was trying to communicate the word *reciprocal*, and he replies, "When she put her hands one way and then reversed them, I knew what she was trying to show."

The teacher was satisfied that the student knew that a reciprocal was formed by exchanging a fraction's numerator and denominator, but wanted to be sure that the students also understood the importance of a reciprocal. He asked another student to use gestures to show the relationship between a reciprocal and its original fraction. Jacob stands and puts up two crossed fingers to represent the multiplication sign, and then immediately after, uses one finger to show the number 1. The teacher asks another student to describe what Jacob is communicating and Selim shares with excitement that, "Jacob was trying to show that multiplying a fraction by its reciprocal is equal to one."

Once the teacher is satisfied that there is productive dialogue about the vocabulary, he introduces the game. The students then immediately try to convince their teammates to choose them as "the communicator." The teacher explains how they will play in teams in their small groups. He looks forward to listening in on the game play as students identify the vocabulary words of the current unit.

Tips from the Classroom

- When students are playing this game in different teams throughout the classroom, it may get rather loud. You may want to suggest that students use a whiteboard and record their guesses, rather than simply calling out. Or, you could move to a larger room, such as the cafeteria, where groups could spread out.
- Encourage students who know the vocabulary word to take a few minutes to formulate a plan with their teammates. This may allow for more thoughtful choices of actions or related words, increasing students' vocabulary knowledge.

What to Look For

- When trying to communicate the vocabulary words, do students attend to the conceptual meanings of the terms?
- What partial understandings or misconceptions do you discover? For example, as students communicate the term *integer*, is it clear that they understand that zero is an integer?
- How precise are students' communication skills? Is there a small group of students who may need to review particular terms?

Variations

- Identify words that the communicator cannot say by writing them on the cards; for example, if the vocabulary word is INTEGER, write *You cannot say POSITIVE or NEGATIVE.*
- Use expressions, numbers, or formulas instead of vocabulary words.
- You can change the time limit of 2 minutes or make a rule limiting the number of guesses that may be made by the guessing team. This change may help some players to think more carefully before making a guess.
- Play the game as you would Pictionary, allowing for drawings to represent the vocabulary word, rather than words or actions.
- Limit students to acting out the words, without saying anything out loud.
- The game can easily be adapted to a competitive one, with larger teams. Teams take turns making a guess, with individual teams receiving points if they guess the word first.

Exit Card Choices

- You are the player who is chosen to communicate the word IRRATIONAL. What words or gestures might you use to encourage your teammates to guess the word?
- Your teammate writes the number *23* on the board. Name three possible terms your teammate might be trying to get you to guess for a vocabulary word.

Extension

You have some classmates who like to play the game *Communicate It!*, but they seem to have trouble getting started when it is their turn. Write possible "starters" for each of the cards in the set.

A-Maz-ing! Game Board

START	-4	$\frac{1}{8}$	0	$\sqrt{25}$	-1.2
3.7	$\sqrt{81}$	1.6	7^2	0.4	$\sqrt{1}$
-3	0.7	$\frac{2}{5}$	-6	$\sqrt{16}$	1.4
$\frac{1}{4}$	5	-1.3	4^2	2.5	$\frac{5}{6}$
5.8	5^2	1.2	$\frac{1}{3}$	16	6
10	$\frac{3}{8}$	$(-4)^2$	-5	6^2	0.8
-3^2	3	-1^2	0.3	-0.4	$\frac{4}{5}$
4	-10	4	$\frac{5}{8}$	$(-2)^2$	FINISH

Why This Game or Puzzle?

Many middle school teachers have shared with us their concern that their students often have difficulty with fraction and decimal computation long after mastery is expected. These same students often resist repetitive lessons on such computational skills, yet willingly play a game which incorporates previously learned content into what they are currently learning. As well as being motivating, there is significant research evidence that active learning is more successful than a more passive approach (Michael 2006).

This game requires students to create equations with either the least or greatest total. By incorporating previously learned skills with fractions and decimals with new sets of numbers such as negative numbers, square numbers, and square roots, students are able to focus on number sense through problem solving, as well as guessing and checking conjectures.

Math Focus

- Computing with various representations of numbers
- Estimating values of numerical expressions

Materials Needed

- 1 *A-Maz-ing!* Game Board per group (page A-18)
- 1 *A-Maz-ing!* Recording Sheet per team (page A-19)
- Chips for placeholders
- 1 six-sided die
- Optional: 1 *A-Maz-ing!* Directions per group (page A-20)

Directions

Goal: Make your way through the maze by creating equations, ending with the least (or greatest) possible total.

- Each team places a chip on the START.
- Teams take turns rolling the die and moving that number of places.
- Players may move their chip to the right or down (or any combination of these two directions), but not left or up.
- After a team has moved its chip the correct number of places, it finds the total of all of the numbers that the chip passed over or landed on while moving through the maze. All operations may be used, and numbers may indicate exponents or be included in scientific notation.
- The team then writes the equation on the recording sheet.
- When one team makes it to the FINISH, the totals are added. The team with the least (or greatest) total wins the game.

How It Looks in the Classroom

Four students in an eighth-grade class are asked to draw a card from the deck that the teacher is holding. The teacher tells these students that they are the number movers for today's warm-up. The four cards chosen are: $\sqrt{81}$, -16, $\frac{5}{8}$, and 6.

The other students in the class are asked to determine an expression that uses all of the numbers to get the least total possible. The students are allowed to work in pairs and may either work mentally or with paper and pencil. Students are given two minutes to determine an expression they think fits the criteria. The number movers are told that they will be busy very soon as they will re-create the expressions at the front of the room.

When time is up, one pair of students immediately states that it wants to go first because it thinks it has the least total. The teacher asks one student in the pair, Reyna, to slowly share her expression. As she does so, the students with the number cards move around at the front of the room. The teacher has also made operations cards for the number movers to hold in their other hands as the expressions are made. Reyna asks the student with the -16 card to move to the left and then says, "Now multiply by $\frac{5}{8}$, then add the square root of 81, and then multiply by 6." The partners believe that they have made an expression equal to -6, but another pair challenges them, and the rest of the class agrees that the answer is actually 44. (See Figure 3.5.) Discussion takes place about order of operations and how parentheses would have been needed if they wanted an answer of -6.

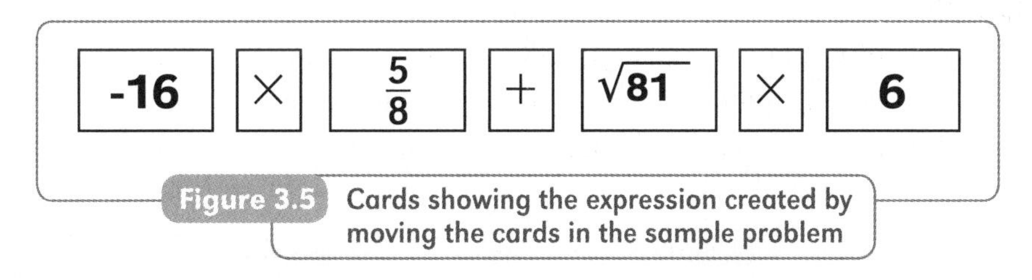

Figure 3.5 Cards showing the expression created by moving the cards in the sample problem

Another pair is sure that it has a lesser total, so the partners report, "Nell should move to the left because she is the square root of 81, then divide by Sam's card which is 6, then multiply by Aaron who represents $\frac{5}{8}$, and then multiply by -16, which is Roshan." After the number movers rearrange themselves, the teacher asks the other students in the class to determine a total, and the partners who created the expression check to see if they are correct. An answer of -15 is agreed upon.

As part of explaining the rules of the game, the teacher asks the students to make thoughtful choices about how they create an expression, just as they did with the introductory activity. Some students remark that trying different ways to make an expression rather than just using the first one that comes to mind would help their game scores.

Tips from the Classroom

- Students on the same team did not always agree on the correct answer to a move through the maze. Players on both teams could be asked to agree on the solution to a move prior to finishing a turn.
- You could choose to have one team challenge the other if they don't think that the answer to the expression is correct. The scores could be altered if the challenge is accepted and the team who created the expression is incorrect.
- When we field-tested this game, we noticed that sometimes there were opportunities for players to make better choices. Encourage students to work together to choose the best moves through the maze prior to making a decision as to which way to move. Ask questions, such as *Is that the best move you can make?* Some students suggested that the other team could earn points for finding moves with higher scores.

What to Look For

- Listen in on the conversations students are having as they play. If you find there are particular misconceptions that several students are having, point them out during a large-group discussion after playing, without naming who had the difficulty.
- Do some students avoid using certain numbers, such as $\sqrt{9}$ —perhaps implying they don't understand how to use them? Encourage students to use as many different types of numbers on the game board as possible.
- Encourage early finishers to search for other paths they could take to meet their goals.

Variations

- Use a calculator to find the equations with the least (or greatest) total.
- Change the available numbers on the game board to include other number sets—for example, irrational numbers written as simplified radicals, such as $3\sqrt{2}$.

Exit Card Choices

- You have the numbers 3, -6, $-\sqrt{64}$, and 0.5 available to use on your turn. How will you arrange these numbers, and what operations will you use, to create the greatest total? the least total?
- Your new game board became smudged after playing several games with it. As a result, there is a number that you cannot read in the series. Name three possible numbers that you think would best fit in the smudged place to find a value of -11. Include the operations you might use as well. The numbers are: -5, 3, ____, -7.

Extension

Create expressions that make a total of 1 to 20 (or -1 to -20) using any numbers you'd like. For each expression, give yourself points for each different type of number that you use (e.g., a negative number, a square root, a fraction, or a decimal).

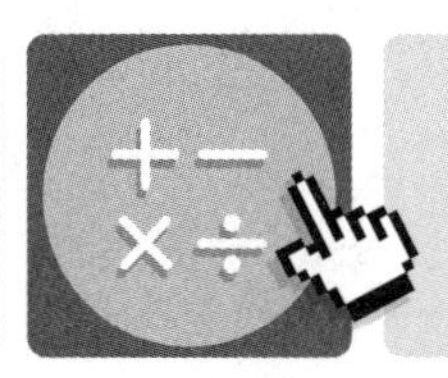

Online Games and Apps

Online games focus on a variety of skills related to our number system, both in terms of equivalence across different representations as well as operations with numbers. For example, there are many games that require players to change a fraction to a decimal equivalent or to divide fractions with accuracy before a bubble bursts, aliens invade, or time runs out. While often engaging for students because of the game themes or the competition, many of these games do not utilize the power of the computer to assist students in learning more deeply about our number system and how to compute with various representations with conceptual understanding.

In contrast, games that provide students visual models of number relationships or engage students in learning that requires perseverance to solve problems can offer a richer learning experience. Choices of difficulty levels also allow for differentiation. Following are some good examples:

- Kakooma, a free game found on Greg Tang's website (http://www.gregtangmath.com), may be played online or through an app. Students may play individually or with others through a multi-player option. The goal of Kakooma, in one version, is to find the number, using positive and negative numbers, that is the sum of two others in a group. Because students must choose the correct sum from among four to nine numbers (depending on the level), flexible thinking must be employed as students consider a variety of choices.

Oftentimes, the player ends up doing a great deal of mental math in order to find the correct answer. There are various ways to increase the difficulty level, by increasing the size of the puzzle or providing a greater target sum. This game encourages efficiency by timing students' completion rates and focuses on accuracy by discouraging students from making incorrect guesses, which result in a penalty time. Competition with other puzzlers can include solvers all over the world.

- Pick-A-Path, a free game from http://illuminations.nctm.org/pickapath, may be played on various devices. Students maneuver Okta, the octopus, through a maze, encountering numbers along the paths. The goal is to create expressions resulting in a maximum, a minimum, or a particular number. If the player does not find the least or greatest number, then he or she is given an opportunity to try again. The score for the round is the total. There are seven levels that may be chosen for this game, each including different sets of numbers in the form of decimals, fractions, negative numbers, and exponents. Students who play this game may be encouraged to team up with a partner, thereby increasing the mathematical discourse as they play, choosing the most appropriate paths.

- Circle 0, a free puzzle from The National Library of Virtual Manipulatives (http://nlvm.usu.edu/en/nav/frames_asid_122_g_3_t_1.html?open=instructions&from=category_g_3_t_1.html), encourages logical reasoning to place positive and negative numbers into intersecting circles. The sum of the numbers found in each circle, once all of the numbers have been placed, must be zero. The puzzle may be solved numerous times as there are many available puzzles. Related puzzles, titled Circle 21 and Circle 99, allow for similar play, but students use whole numbers or decimals to complete them.

When playing online games, students should still be expected to provide evidence of their learning. For example, if playing Pick-A-Path, students may be asked to record their moves in order to demonstrate how they found the answer and to allow the teacher to assess their thinking once the game is complete. You may also ask students to answer questions about what they learned while playing a game, such as asking them to write about the strategy they used to complete a Circle 0 puzzle.

Ratios and Proportional Relationships

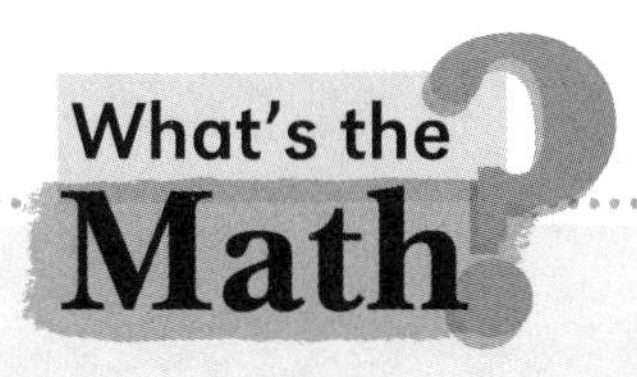

Understanding the concepts involved in ratios and proportional reasoning is essential to continued study in both mathematics and the sciences (Common Core Standards Writing Team 2011a). In math, proportional thinking underpins such ideas as slope, similar figures, and trigonometric functions, while in science examples include acceleration, density, and equilibrium. Ratios and proportional relationships also pervade our daily lives, such as when we shop, cook, measure, or listen to the news or weather. It's clear, therefore, that students must develop conceptual understanding and proficiency in these topics.

A ratio is used to relate two or more quantities. The quantities may have the same units, such as cups of oil to cups of vinegar, or they may have different units, such as miles to hours. There are many ways to express ratios—for instance, four students for every one adult, four times as many students as adults, four students per one adult, or four students to one adult. The phrase, *times as many as*, can help students to remember the multiplicative relationship between the quantities, which is a challenging concept (Lamon 2012). Students can find equivalent ratios through additive thinking, but it is not an efficient method when the gap between numbers widens.

Two students' responses to the question *If the team has 3 gloves for every 6 baseballs, how many baseballs will the team have for 9 gloves?* are shown in Figure 4.1. Note that, though both students arrived at the correct answer, Student A clearly understood the idea of what it means to have 3 gloves for every 6 baseballs but used additive thinking, while Student B demonstrated multiplicative thinking. It is tempting to simply mark a student's work as correct if the answer is accurate, which would occur in both instances with these exit cards. However, when choosing to look at work more closely, we have the opportunity to note the methods students use and to follow up with a question or discussion to stimulate deeper conceptual understanding or re-evaluation of the response. How might you respond to Student A? For example, would you ask how he or she might use multiplication here? Would you ask Student B to better differentiate the multiplication sign from an addition sign?

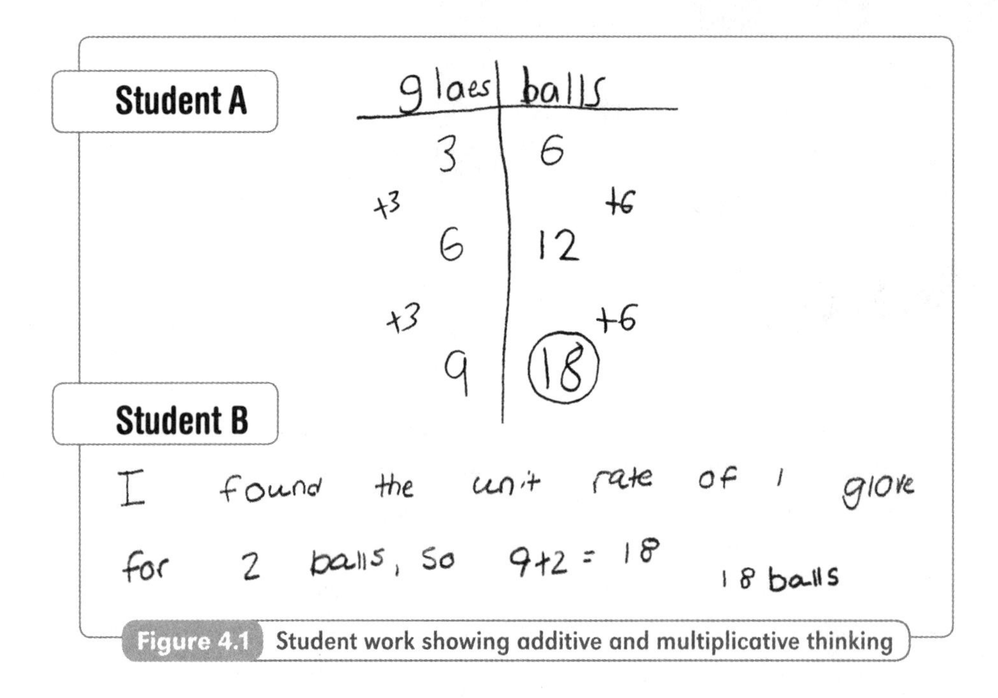

Figure 4.1 Student work showing additive and multiplicative thinking

You could choose to use the notation 1:2 or $\frac{1}{2}$ to represent this ratio. There are also a variety of visual representations that you could use to represent this relationship, including a table, a bar diagram (also known as a strip or tape diagram), a double number line, and a graph on the coordinate plane. Such models help students to identify equivalent ratios and to compare ratios (Collins and Dacey 2014).

In sixth grade, students work on developing a conceptual understanding of a ratio and learn the vocabulary to describe the relationship. They are expected to use a variety of visual representations of ratios and to use ratios to solve real-world problems. Students are also introduced to unit rates at this level, as well as the idea that a percent is a rate per 100. Students' understanding continues to deepen in seventh grade as they extend their understanding of ratio to proportional reasoning. These students need time to explore a variety of strategies, including making tables, diagrams, and graphs, before they eventually use equations to represent proportions. The variety of real-world contexts to which seventh graders apply their understanding also broadens, including, for example, situations involving percent increase or decrease. Eighth graders must maintain these understandings and skills to apply them to algebraic thinking and to solving real-world problems.

When playing games or solving puzzles, it is often best to make estimates first. Players and puzzlers then need to decide whether mental arithmetic, paper and pencil, or perhaps calculators are best used to find exact answers. Through this process students gain a better understanding of when to use various approaches, which will serve them well in real-world situations.

Four of a Kind Card Set A

$6\frac{1}{2}$ to 4	13, 39, 117 / 8, 24, 72
$\frac{26}{16}$	
5 for every 3	15 to 9
	10, 20, 30 / 6, 12, 18
20, 30, 40 / 24, 36, 48	$\frac{10}{12}$
	15:18
$\frac{4}{5}$	
36 per 45	8, 12, 16 / 10, 15, 20

Why This Game or Puzzle?

According to Margaret Rathouz and her coauthors, "The development of precise multiplicative language is critical in understanding ratio relationships in preparation for later work with rates, scaling, proportions, and the concept of slope," (2014, 40). Such language includes a variety of phrases, such as *for each*, *for every*, *times as many as*, and *per*. We want our students to use this vocabulary as well as to connect it with other non-verbal representations of ratios. Students should be able to explain, for instance, how *three dollars for each box of colored pencils* would be represented in a table, as well as how they could look at the associated values in a table and identify this relationship.

We've taken a game we learned from Michael Schiro (2009) and adapted it to emphasize the important links among the various ways in which ratios can be represented. The goal of the game is to get four cards with equivalent ratios. The ratios are represented with words, pictures, symbols, tables, and double number lines. Two versions of the card set are provided; Set A involves simpler ratio equivalents than does Set B.

Math Focus

- Recognizing equivalent ratios shown in a variety of representations
- Using ratio language to describe a relationship between two quantities

Materials Needed

- 1 *Four of a Kind* Card Set A or B per group (page A-21 or A-22)
- Optional: 1 *Four of a Kind* Directions per group (page A-23)

Directions

Goal: Be the first to collect four playing cards with equivalent ratios.

- Shuffle the cards.
- Deal four cards to each player.
- At the same time, each player looks at his or her own cards and decides on one card the player does not want. The player places that card facedown in front of the player to the right.
- All the players pick up their new cards, at the same time, so that each person once again has four cards in his or her hand.
- Players continue to pass and pick up cards, waiting for all players to pick up before the next pass begins.
- The first player to get four cards with equivalent ratios says, "Four of a Kind!" and wins.

How It Looks in the Classroom

One sixth-grade teacher draws a double number line diagram on the board (see Figure 4.2) and tells students to talk to others at their table about different ways to represent the relationship shown. As the teacher circulates, she recognizes that some students are more at ease than others with finding ratios that share this relationship. Some students draw pictures of objects representing the first two numbers in the diagram and yet seem uncertain as to how to write the simplified fraction form for the ratio. The teacher asks several students to show their ratios on the board next to the diagram; she includes a student who has written the correct numbers but in reverse order (7 to 10), for she is confident that this representation may be mistakenly chosen by other students as well.

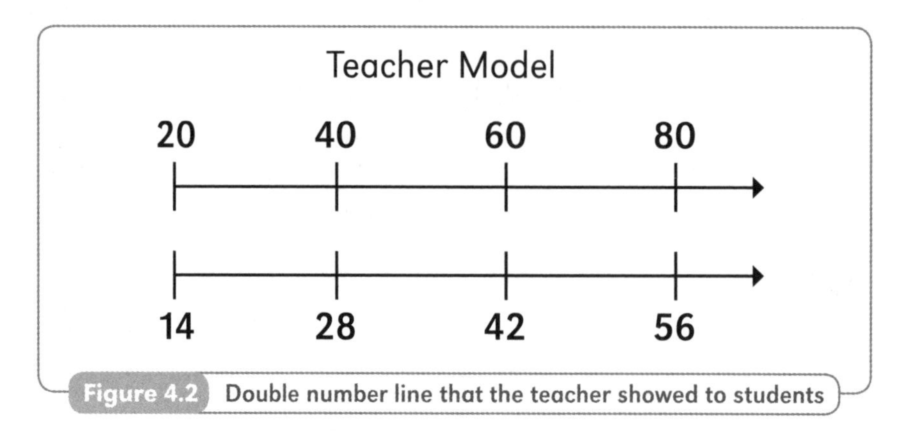

Figure 4.2 Double number line that the teacher showed to students

The teacher asks students who contributed a response to explain their representations. All students agree that Sophie's representation of $\frac{20}{14}$ is accurate, as this is the fraction representation for the first two numbers of the double number line diagram. When Edmar explains his answer, writing *30 to 21*, some students question him and ask why he didn't add 10 to the 14 to get 24, after he added 10 to the 20 to get 30. Yet another student asks why the first number would not be 27, adding 7 to the 20, as Edmar did to get from 14 to 21. While the teacher knows that she has provided direct instruction related to finding ratios in recent lessons, she also knows that understanding how additive thinking can and cannot be applied to finding equivalent ratios is not always easy.

Rather than stepping in to give her own explanation, the teacher is eager to facilitate a class discussion among her students. She knows that there are opportunities for powerful learning when students listen to their classmates provide explanations and then reach their own conclusions. She asks for a volunteer to give an explanation for Edmar's answer and Hee-Young is eager to share how her visual representation can be helpful to her classmates. She shows how 10 triangles and 7 rectangles represent the same ratio as $\frac{20}{14}$ because, when you add another 10 triangles, you must add another 7 rectangles. She explains, "This is like adding another set of what I have in my picture and if we add one more set, we get 30 to 21." (See Figure 4.3.)

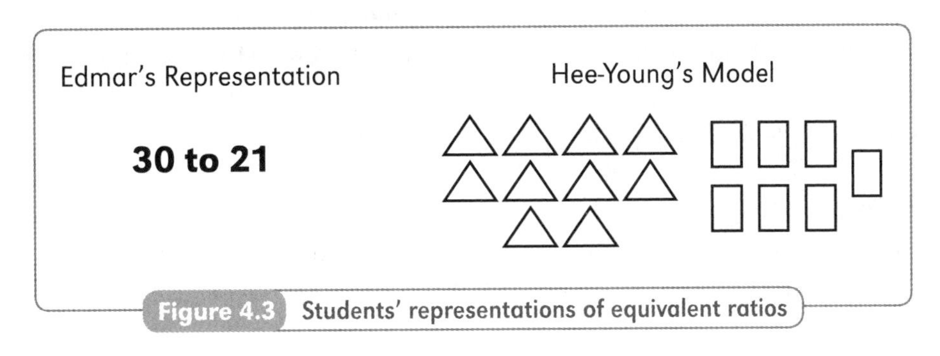

Figure 4.3 Students' representations of equivalent ratios

The teacher introduces the game to the students, emphasizing that playing the game will give them more opportunities to practice finding equivalent ratios using different representations. As play begins, a student is overheard saying, "I think that I'm going to need to talk to you about these ratios as we play the game."

Tips from the Classroom

- To keep the timing of the card passing in sync, you may want to have a player in each group say "pass" and "pick up" at an agreed-upon time.
- Some players enjoy the speed involved in a competitive game such as this one. If you want the players to slow down some, consider introducing the cooperative version described later in the "Variations" section.

What to Look For

- Do students recognize the ratio in one form more easily than in another?
- How do students talk about a ratio that is equivalent to another?
- Do players recognize equivalent ratios in their hand?
- What evidence do you see that suggests students are engaged in multiplicative thinking?

Variations

- Differentiate, as needed, by choosing which cards are used in the game. In initial play, you may wish to eliminate some of the ratio forms, or have students use only cards with smaller numbers.
- For each ratio, create a card showing a coordinate graph representation of the ratio and change the goal of the game to require five of a kind to win.
- Have students play cooperatively. Hands of cards stay faceup for all to see throughout the game. There is no talking allowed in the game but players can gesture to each other. The goal is to have everyone get a set of matched cards in the fewest number of passes.

Exit Card Choices

- What representation would you use to help you find an equivalent ratio for 45:16? Why?
- How would you help someone understand that 15 for every 4 is not equivalent to 20 for every 9?
- Create a ratio table to show equivalent ratios for 6 to 5.

Extension

Have students make posters for display in your classroom that depict real-life situations in which ratios are used, and have them include different representations for the ratios.

Best Unit Price

Why This Game or Puzzle?

Unit price labels can be confusing. According to *Consumer Reports* (2015), eight in ten Americans rely on unit price labels to make economic decisions when shopping, yet many such labels are inconsistent. For example, the organization found that some labels for salad dressing were priced per pint, while others were priced per quart. Consumers need to develop the ability to compare unit rates while shopping and may find estimation more efficient than looking for small, sometimes inconsistent labels.

In this game, teams are dealt cards with a variety of information about products—for example, a card might describe a bottle of Marie's Moisturizing Shampoo containing 12 fl. oz. with a price of \$8.00. Among the products, larger sizes do not always have a better unit price. There are four cards in the deck for this shampoo, each listing a different volume and price. If the top card of the deck is turned over and reveals a card about this shampoo, either team with a shampoo card must show it. The unit rates are then compared. If a team's card shows the best rate, that team keeps the shampoo cards; if the deck's card shows the best rate, the shampoo cards are placed in the discard pile. Logical reasoning may often be used in lieu of exact rates. For example, to compare the card for Marie's Shampoo described above to a card that shows a volume of 24 fl. oz. and a price of \$15.50, a player might think, "24 fl. oz. is twice as much as 12 fl. oz., but \$15.50 is less than two times \$8.00, so that bottle has the better unit price."

***Best Unit Price* Cards**

Marie's Moisturizing Shampoo 1 pt. 2 oz. \$9.00	Marie's Moisturizing Shampoo 20 fl. oz. \$11.00	Marie's Moisturizing Shampoo 24 fl. oz. \$15.50	Marie's Moisturizing Shampoo 12 fl. oz. \$8.00
Fine Line Black Pens Box of 100 \$89	Fine Line Black Pens Pack of 10 \$12.99	Fine Line Black Pens Pack of 5 \$7.50	Fine Line Black Pens Pack of 2 \$3.50
$\frac{3}{8}$"-wide Ribbon 12 yards \$3.25	$\frac{3}{8}$"-wide Ribbon 27 feet \$2.75	$\frac{3}{8}$"-wide Ribbon 7 yards \$2.50	$\frac{3}{8}$"-wide Ribbon 20 feet \$2.50
Seedless Green Grapes 1 lb. 6 oz. \$3.25	Seedless Green Grapes 2 lb. 10 oz. \$7.00	Seedless Green Grapes 24 oz. \$4.50	Seedless Green Grapes 18 oz. \$3.60

Math Focus

- Using ratio and rate reasoning to solve real-world problems
- Finding unit rates
- Using ratio reasoning to convert measurement units

Materials Needed

- 1 deck of *Best Unit Price* Cards per group (pages A-24 to A-26)
- Optional: 1 *Best Unit Price* Directions per group (page A-27)

Directions

Goal: Collect the most cards.

- Shuffle the cards and deal each team six cards, which the teams keep private.
- Place the remaining cards facedown as a deck.
- On each turn:
  1. Turn over the top card of the deck.
  2. Any team that has a card in its hand with the same product as the card turned over puts that card faceup in front of its players.
  3. Players compare to see which card shows the best unit price.

If a team has the card, it puts these cards in its pile of winning sets. If the deck's card shows the best unit price, all the played cards are placed facedown in a discard pile.

4. Teams take a card from the top of the deck to replace any cards played, keeping six cards in their hand as long as possible.
5. If neither team has a card with the same product as the deck's card, that card is placed in the discard pile.

- When all of the cards in the deck have been turned over once, teams take turns putting down a card. If the other team has a card with the same product, the unit prices are compared. If the other team does not, the card is added to the discard pile.
- When teams no longer have cards with matching products, the cards in the winning sets are counted and the team with the greater number of cards wins.

How It Looks in the Classroom

Four different-size ketchup bottles sit on a table in the front of one seventh-grade class. Next to each bottle is a card with the price for the ketchup. The teacher asks his students which one they would buy. Kit comments that she would buy the largest bottle because they use a lot of ketchup in their house. Jordan says that he'd buy the medium-size bottle because that's the one his dad says best fits in the refrigerator, and Dominic adds that he learned to buy the one that is the best price. His friend leans over and responds, "Well, that's always the cheapest one, isn't it?" The teacher interjects at this point, asking the class, "What does the cheapest price mean to you?"

This question opens a discussion of different ways to think about what makes a bottle of ketchup the cheapest, and eventually leads to students thinking about how comparing the prices of the ketchup bottles also involves thinking about the measure of each bottle. The concept of unit rate arises naturally from this discussion and the students calculate the cost of the ketchup in each bottle per ounce. The teacher listens in on the groups' conversations, noting the students who are unsure how to find the unit rates due to the various units of measurement written on the bottles or the computation involved in the task.

As the students come to the conclusion that the third-largest bottle is the best price, a student who thought that the largest bottle was always the best deal remarks that, from now on, she is going to be a more careful shopper.

The teacher describes the game the students are about to play, telling them it is quite similar to the ketchup task. After he reviews the rules of *Best Unit Price*, the students engage quickly in choosing a dealer and have paper and pencil ready when needed.

Tips from the Classroom

- We found that some students automatically computed each unit rate, even though deductive reasoning could have been a more effective way to make some of the comparisons. You may want to encourage students to consider alternative approaches.
- Some students would benefit from creating a conversion chart to support them while playing the game.

What to Look For

- Do students convert different units of measurement as they find unit rates?
- What strategies do students use to determine the best price? Do they always find the unit rate, or do they use other methods for comparison as well?
- What visual representations, if any, do students use to help them make comparisons?
- What efficient strategies do you observe that you would like students to share with their classmates?

Variations

- Some students suggested that the deck could be used only for getting more cards, not for initiating comparisons. In such a scenario, students thought a player could begin a turn by saying something like, "Do you have a card about shampoo?" You may ask your students to further articulate the directions for such a game and then play it.
- Students could continue to play by shuffling and reusing the cards in the discard pile, ending the game only when no further matches are possible.

Exit Card Choices

- Is 12 oz. of mustard for $5.98 or 9 oz. of the same mustard for $4.25 a better unit price? How do you know?
- When comparing the prices and quantities of two items in a store, how do you know which one is a better financial deal?
- While playing the game, how did estimation help you to compare the costs of the products? Give two specific examples.

Extension

Have students explore online store flyers for examples of inconsistent labels or to find best buys among local stores. Create a display within your classroom that can be added to throughout the school year.

Why This Game or Puzzle?

Rates are often challenging to students, as they frequently involve the need to coordinate two different units, such as *miles* per *hour* (Lobato, Ellis, and Charles 2010). Also, we often need to convert from one unit of measure to another when solving problems—for instance, *At a rate of 12 miles per 15 minutes, how far can you travel in 1 hour?*

Keep Going? is built on the game of blackjack in that teams want to reach a total distance as close to 21 miles as possible, without exceeding that mileage. Cards in the game list an activity, a speed, and a distance. As in blackjack, students are dealt two cards and must then determine whether or not to accept additional cards.

***Keep Going?* Cards**

Skateboard 40 minutes 7.5 mph	Rollerblade 20 minutes 10 mph	Walk 40 minutes 3.5 mph	Run 30 minutes 6 mph
Jog 75 minutes 4 mph	Bike 20 minutes 12 mph	Unicycle 15 minutes 2 mph	Car 5 minutes 60 mph
Skateboard 30 minutes 12 mph	Rollerblade 40 minutes 10 mph	Walk 60 minutes 4 mph	Run 30 minutes 5 mph
Jog 45 minutes 4 mph	Bike 24 minutes 15 mph	Unicycle 30 minutes 3 mph	Car 10 minutes 42 mph

Math Focus

- Solving unit rate problems involving constant speed
- Using ratios to convert measurement units

Materials Needed

- 1 deck of *Keep Going?* Cards per group (page A-28)
- 1 *Keep Going?* Recording Sheet per team (page A-29)
- Optional: 1 *Keep Going?* Directions per group (page A-30)

Directions

Goal: Get a total distance closest to 21 miles without going over.

- Choose a player to also serve as the dealer. The dealer shuffles the cards and deals two cards to each team with one card faceup and one card facedown.
- Teams may look at the facedown card, but keep it private from the other team.
- Place the remaining cards facedown as a deck.

- On each turn:
  1. Each team may choose to "stop" (not take any more cards) or to "keep going" (request another card from the dealer).
  2. If a team chooses to "keep going" by receiving another card, the dealer turns over two cards from the deck and the team chooses one of the cards. The card not chosen is returned to the bottom of the deck.
  3. The team determines how close it is to 21 miles and records its mileage on the recording sheet.
  4. The team continues steps 2–3 until it chooses to stop.
- Once both teams have had their turn, they each reveal how close they are to 21 miles.
- The team that is the closest to 21 miles, without going farther, is the winner.
- A new dealer is chosen for the next game and all cards are then returned to the deck and shuffled, and play begins again.
- Play continues for a predetermined period of time.

How It Looks in the Classroom

The teacher decides to introduce the *Keep Going?* game by asking his students if they know what a triathlon is. A number of students comment that they know older siblings or other relatives who have participated in one, and John describes his understanding of a triathlon as riding a bike, swimming, and running to win a race. The teacher tells John that his description is a good one and that they are going to play a game, similar to a triathlon, involving different ways to travel.

The teacher introduces this game by giving each group of students six cards (Figure 4.4) and asking them to choose cards from the set that they think will get them closest to 5 miles without going over. One group decides that its members will work together to find the miles traveled for each of the cards, while another group splits up the work to find the mileages, with each group member choosing one card to consider and then members who finish quickly choosing a second card. When the teacher asks groups to report their findings, the students in Park's group are certain they are the closest and he shares that they reached $4\frac{1}{2}$ miles using cards A, B, C, and F. The teacher asks them to explain how they found their answer and they show their representations to indicate how they solved for the number of miles for each form of travel.

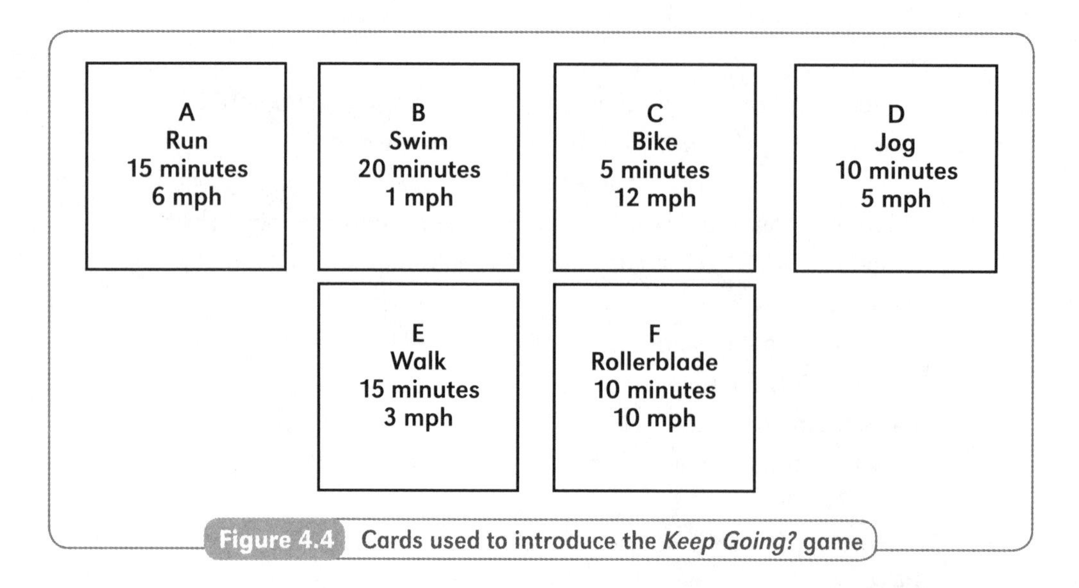

Figure 4.4 Cards used to introduce the *Keep Going?* game

As other groups share their solutions, one group thinks that it is closer to 5 miles, but then realizes that it is over the goal. Some students talk about how fast they could swim or bike; the teacher is encouraged to hear how engaged they are with the context of the game.

The teacher explains that the game students will play in their groups is similar to what they just did, but with a new twist: they will be given only two cards to start and then can choose more cards, one at a time—but they won't get to see all of the cards at once. The students acknowledge that this will be more challenging than the last activity, but they are up to the task.

Tips from the Classroom

- Some students may need more support when playing the game and may benefit from making a conversion chart.
- When we field-tested this game, some players did the work mentally while others needed a more structured recording sheet to record their steps. You may want to adapt the recording sheet to meet the needs of your students.
- Some students may benefit from your encouragement to stop and think about the data on a card before immediately beginning to compute. As Stacey proclaimed, "Wow, thinking about this first really helped me. I know I can't take this card. This distance will be way too far."

What to Look For

- Do students check each other's work to be sure that the computation is correct?
- Do students accurately compute the number of miles traveled?
- Do students sometimes estimate the distance indicated on a card to determine whether finding the exact value would be worth their while?

Variations

- Change the goal to 21 kilometers rather than 21 miles and use all metric measures.
- Create cards that include both metric and non-metric measures within the same set.

Exit Card Choices

- Who went farther: you, who skateboarded at a rate of 7 mph for 15 minutes, or your friend, who ran at a rate of 6 mph for 20 minutes? Explain how you know.
- You are running a 5-km race and you have set a goal to finish within 30 minutes. At the 1-km marker you have been running for 8 minutes. If you keep this pace, will you meet your goal? Explain your thinking.

Extension

Bring students to an open space and have them determine their own speeds for various types of travel. Students could include those types listed in the game cards as well as other forms of travel, such as walking backward or hopping. Once data are collected, challenge students to combine at least five of their modes of travel and the amount of time for each, to reach a total as close as possible to, but not farther than, 21 miles.

The Question Is/The Answer Is

Why This Game or Puzzle?

We all know the difficulties students experience when solving a proportion, especially for those who have been taught to cross multiply. Too often, students use this technique when a proportional relationship does not exist or misapply the rote approach by not keeping track of parts and wholes appropriately. Use of bar diagrams helps students to identify the proportional relationships within a problem (Cohen 2013). Encourage students to use these diagrams or double number lines to establish the relationships among the quantities.

The Question Is/The Answer Is Cards

The Numerical Answer Is 112 *The Question Is* Kai's birthday party will cost $120 if he invites 10 guests. If 2 more guests come to the party, how much will it cost for Kai's party?	*The Numerical Answer Is* 144 *The Question Is* If it costs you $55.80 to fill your 20-gallon gas tank, but you only have $34, how much gas (to the nearest gallon) could you buy?
The Numerical Answer Is 144 *The Question Is* A store has T-shirts on sale at 3 for $16.50. If a customer spent $66 on T-shirts, how many did he buy?	*The Numerical Answer Is* 12 *The Question Is* The Garcia family traveled 273 miles in 4.2 hours on the first part of their vacation. If they traveled at the same rate to get to their next destination, how far would they travel in 6 hours?
The Numerical Answer Is 12 *The Question Is* You invest $3,600 in a stock, for one year, that pays a $162 dividend. At the same rate, how much will you need to invest to earn $270?	*The Numerical Answer Is* 390 *The Question Is* The student council representatives determine that they can raise $85.80 if 13 people buy water bottles. How much money will they raise if 72 people buy water bottles?
The Numerical Answer Is 6,000 *The Question Is* Marcus works 15 hours mowing lawns and makes $216. How much money will he make if he works 10 hours?	*The Numerical Answer Is* 475.20 *The Question Is* If an airplane can travel 2,400 miles in 6 hours, how far can it travel in 15 hours?

Math Focus

- Representing proportional relationships
- Using proportional relationships to solve real-world problems

Materials Needed

- 1 deck of *The Question Is/The Answer Is* Cards per group (pages A-31–A-32)
- Optional: 1 *The Question Is/The Answer Is* Directions per group (page A-34)

Directions

Goal: Place cards so that the answer identified on each card answers the question on the card before it.

- Spread out the cards faceup on the table or the floor.
- Choose a card and read its question.
- Find a card with a matching answer, place this card next to the first card, and read the question on this second card.
- Continue to read questions and find answers. Organize the cards in a circle so that each question is followed with a correct answer.
- Each card must be included in the circle.

How It Looks in the Classroom

One teacher introduces the class to *The Question Is/The Answer Is* by having her students solve a similar puzzle as a whole group. She explains that the puzzle cards must be placed in order so that the question on each card matches with the answer on the adjacent card, including the first and last card. Each of the puzzle cards in the mini-puzzle (Figure 4.5) is displayed on the whiteboard so that the cards can be moved around easily.

A
The Numerical Answer Is
245

The Question Is
Casa babysat for 10 hours last week and earned $200. She will babysit for 12 hours this week. How much will she make babysitting for 12 hours?

B
The Numerical Answer Is
240

The Question Is
You buy a new bike, which is on sale. You end up paying $150 after a 25% discount. What was the original price of the bike?

C
The Numerical Answer Is
240

The Question Is
A fishing boat cruises for 4 hours traveling 140 miles. At the same rate, how many miles can the boat travel in 7 hours?

D
The Numerical Answer Is
200

The Question Is
You go out to dinner for your birthday. You want to give a 20% tip. If the bill was $200, what total amount will you pay?

Figure 4.5 Mini–*The Question Is/The Answer Is* puzzle

Derek comments to his teacher, "You always make us put labels on our answers when we are doing a real-life problem. You didn't do that in this puzzle." The teacher is pleased that Derek has noticed the absence of labels, as she has emphasized the need to use them whenever the answer represents an amount of something. She replies, "In this case I haven't put labels on the answers because I don't want you to just match labels as a way to solve the problem. So, I only want you to find the numerical part of the answer. But when you talk in your groups, I would like for you to include them."

Amar suggests that the card about the bike should be placed next to the card about the babysitter, but Fabia thinks that's the wrong answer. The teacher is pleased when Fabia asks if she can show the class how she got her answer to the bike problem, as she has not always felt comfortable sharing in front of the class. Fabia told her teacher a few days earlier that she liked using a bar diagram to visualize how to solve such problems. She shows the class the diagram she made for this problem (Figure 4.6) and explains that she matched the 150 and the 75% and then labeled the numbers and the percentages in such a way that she could evenly distribute them across the top and bottom of the diagram. She concluded that the regular price of the bike must have been $200 because 100% represents what the bike cost originally. The class agrees that Fabia is correct in her thinking, and several of her classmates tell Fabia that they solved it that way too. The card about the bike is then placed to the left of the card about the dinner and the students complete the mini-puzzle as a group.

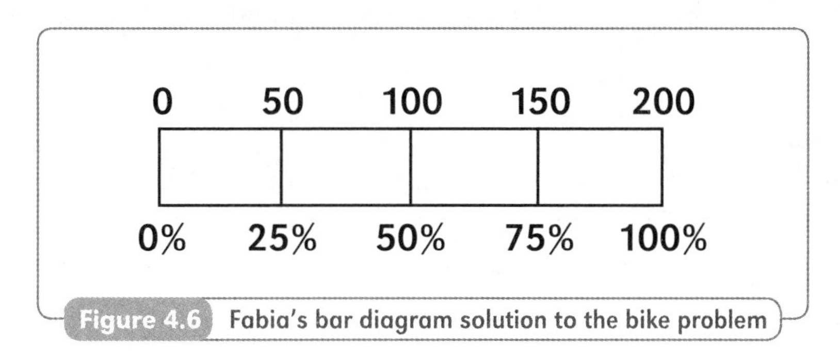

Figure 4.6 Fabia's bar diagram solution to the bike problem

The teacher then distributes the cards for the puzzle for students to solve in groups. Today she is giving the students a choice as to which partners they will work with, and she notices that several students who earlier discussed their similar strategies have asked to work together. She has decided not to remind the students that there are some cards that have the same answer, yet only one placement of cards will allow for all of the cards to be used. She would rather have the students work through this challenge when they come across it in the puzzle.

Tips from the Classroom

- Putting tape on the back of the cards to hold them in place may help some students to organize their work more easily.

- Some students may find it easier to place the cards in a line. Just remind them that the last card must lead back to the first.
- As with many puzzles (and real-life problems), sometimes the best thing to do when "stuck" is to take a break. Have students take a picture of their work thus far, so that they can re-create it easily.
- Some students wanted to write on the cards to keep track of how they were solving some of the problems. Suggest ahead of time that students have paper and pencil ready for such recordings.
- Students can be assigned to other groups in the role of advisor or checker once they have found a correct solution. Encourage these students to ask and respond to questions, rather than supplying the group with a correct answer.

What to Look For

- How do groups decide how to begin solving the puzzle?
- How do students decide to solve the problems? Do they change the given data to a unit rate or set up and solve a proportion?
- What models, such as bar diagrams, do students use to help them understand and apply the given proportional relationships?
- What do students do when they discover there is more than one card with the same answer?
- What do students do when they get "stuck" and can't find a way to fit all of the cards together?

Variations

- Add an "ask a friend" rule, which would allow students to ask a classmate working in another group for a hint or to verify whether an answer is correct.
- Vary the level of difficulty by asking some students to solve the puzzle with paper and pencil and others with a calculator.

Exit Card Choices

- Which question did you find most challenging to solve? Why?
- Write two different questions for which the numerical answer is 150. The labels for the answers must be different as well.
- A large Douglas fir tree has a height of 144.5 feet and a circumference of 17 feet. If a small Douglas fir has the same ratio with a height of 50 feet, what is the circumference of the tree, to the nearest inch?

Extension

Have students create their own puzzles using the template on page A-33. The easiest way to do so is to start with the question on the first card and then proceed from answer to question until the end, when the answer to the last question is recorded on the first card. Once recorded in this order, any card may be shown randomly as the first card.

Money Matters Puzzle A

Name(s): ____________________ Date: __________

Lewis, Marie, Norah, Odette, and Portia each went shopping. They each spent a different amount of money and a different percentage of the money they brought shopping. The amounts of money they spent were $25, $36, $54, $70, and $100. They spent 12.5%, 25%, 50%, 75%, and 90% of their shopping money. Use these clues to find how much money each person had for shopping.

- Lewis had $80 more shopping money than Norah.
- Norah and Portia both had the same amount of shopping money.
- Marie had less than $50 of shopping money.
- Odette spent three-fourths of her shopping money.
- Portia spent four times as much as Norah spent.

	Amount of Money Spent					Percent of Money Spent				
Names	$25	$36	$54	$70	$100	12.5%	25%	50%	75%	90%
Lewis										
Marie										
Norah										
Odette										
Portia										

Lewis had $__________ for shopping.

Marie had $__________ for shopping.

Norah had $__________ for shopping.

Odette had $__________ for shopping.

Portia had $__________ for shopping.

Why This Game or Puzzle?

In real life, information may be stated implicitly, rather than explicitly, requiring the ability to use deductive reasoning. Also, information may be redundant or need to be combined to be most useful. Too often, we give students mathematical tasks that provide all of the information explicitly, without extra data, and already in the order that it is needed for use.

Money Matters puzzles require solvers to interpret and combine clues to identify the solution. They also provide the opportunity for students to learn, if they do not already know, how a table can be used to organize information. Visual organization of data can support students' conceptual understanding (Farbermann and Musina 2004), and it is important to expose students to a variety of visual formats.

Students at this level are expected to solve realistic percent problems involving taxes, tips, commissions, increases, and decreases, though these topics are sometimes less directly relevant to the lives of middle school students. Logic puzzles can provide a motivating context in which to pursue such examples. Two levels of the puzzle are provided; Puzzle A focuses on finding the whole when given the part and the percentage, and Puzzle B on finding the part when given the percent of increase and the whole. Note that some students solving the latter may incorrectly believe that they can simply subtract the percent increase from the current summer's pay, rather than setting up a correct proportion.

Math Focus

- Representing proportional relationships between two rates
- Using proportional relationships to solve real-world problems involving ratio and percent

Materials Needed

- 1 *Money Matters* Puzzle A or B per student or pair (page A-35 or A-36)
- Optional: 1 *Money Matters* Directions per student or pair (page A-37)
- Optional: 1 calculator per student or pair

Directions

Goal: Use clues to match the right pieces of information (for example, a person's name with the money he or she spent and a person's name with the total amount of money he or she borrowed).

- Work alone or with a partner.
- Read the clues.
- Use the table to organize what you know from each clue.
- Make notes so that you can recall your thinking. Include computation and clue numbers in your notes so that you can refer back to them if you need to.
- You do not need to use the clues in order.
- When you are finished, check your solution with each clue to be sure that it works.

How It Looks in the Classroom

Students in this seventh-grade class have solved logic puzzles before but may not be familiar with solving them when they contain clues related to mathematics. The teacher reminds the students that their goal is to solve the puzzle she has displayed (Figure 4.7) by obtaining information from the given clues and filling in the table as they find solutions. The teacher asks students to talk at their tables about how they should begin. Saul tells others at his table that he would find all of the possible discounts and the new prices. Evan says that even though there aren't that many combinations, he thinks that there should be an easier way.

Tom's TVs is having a sale. Determine which discount is given for each TV. The original TV prices are $350, $425, and $500. The discounts are 20%, 25%, and 35%. Each percent discount is matched with a different-priced TV.

1. The amount of money discounted off the price of the $425 TV is more than that of the $500 TV.
2. One of the TVs cost $318.75 after the discount was applied.

Original Prices	20%	25%	35%
$350			
$425			
$500			

Figure 4.7 Mini-puzzle used to introduce *Money Matters*

Hana is asked to summarize her group's conversation, and she starts by saying, "A couple of us wanted to find all the combinations first. But Finn suggested that we use the first clue to consider the relationship between the possible discounts and the original prices of those TVs using some mental math. Sam figured out that it couldn't be 20% off of $425 or $500 because that would make the $500 TV's discount more than the $425 TV's." Sam is eager to share that she used the mental math strategies that she had learned earlier in the week to figure out the discount amounts, first finding 10% for each discount and then multiplying as necessary. The teacher reminds the students to put an *X* in the table when they know that a particular combination does not work.

The students continue to share different strategies for solving the puzzle and are reminded, once it is solved, to check their solution to be sure that it works with the given clues. As the students are given the first *Money Matters* puzzle to solve, they are challenged to find as many answers as possible using strategies other than finding all combinations to solve it.

Tips from the Classroom

- Some students may find it easier to organize information in a list. For example, they could write all possible percentages of money spent under each person's name and then cross out percentages they eliminate and circle percentages they identify.
- Students should be encouraged to eliminate as many possibilities as they can through deductive reasoning and then find exact answers to check their final conclusions. Making calculators available may allow some students to give greater attention to the proportional, as well as the deductive, reasoning required to solve the puzzle.
- In our field-testing, some teachers were concerned that the puzzles might be too challenging, but were surprised by what students could understand when they worked together.
- If necessary, you could highlight a clue for some students and suggest that they consider that clue first.
- Be sure students use pencils, or set up an online document to be reused, so that solvers can begin again when necessary. Alternatively, students can use multiple copies of the chart so that they can refer back to previous work if they choose to do so.
- Students may want to use an online educational app that allows them to chat with others about their current thinking in solving the puzzle.

What to Look For

- Is students' proportional reasoning sound?
- Are there vocabulary words in the puzzle that students find challenging or confusing?
- Do students use the clues in an efficient order?
- Are students drawing conclusions from implicit information? For example, in the first puzzle, one of the clues is *Portia spent four times as much as Norah spent.* Do students recognize that this clue, combined with the clue that states they had the same amount of shopping money, indicates a relationship between their percentages? You should emphasize this type of thinking when the class debriefs its solution strategies.

- How do students record the information they are learning from the clues?
- Do students check their solutions?
- How successful are students in describing their solution paths to others?

Variations

- Have students complete the puzzle in groups of five, giving each group member only one of the clues.
- Do not give students access to the chart, requiring them to decide for themselves how best to organize the information.

Exit Card Choices

- Change one of the clues in Puzzle A so that it is more challenging, but still provides the same information.
- Which clue did you think/talk the most about? Why?
- Which clue was the most useful to you? Why?

A student's response to this third exit card question is shown in Figure 4.8. The student has clearly recognized the relationship between the two clues about these girls. He may or may not have deduced the inverse relationship between their spending money and the percentages they spent. What might you ask to further probe his thinking? The response does communicate his excitement at finding this connection—something always pleasing to see.

Portia spent 4 times as much as Norah spent is the most useful Clue because it was the first clue that I Knew I was going in the right direction, I know that they spent the same amount so then I looked at the percents and 12.5 x 4 = 50 and that was the only one that worked, AWESOME !! Then it was easy to see that Nora spent $25 and Portia spent $100. 25 x 4 = 100

Figure 4.8 Student work for the third exit card

Extension

Give students the answers to a puzzle and have them work backward by writing the clues. Have students share the clues with each other.

Online Games and Apps

The games included here are examples of how technology can support students' learning of ratio and proportion concepts. Each game provides students with the opportunity to visualize a ratio in a variety of contexts, as well as to freely manipulate the environment in order to test out theories until the concept is well understood. Two of the games also engage students in a real-world environment of interest to many middle school students, with a focus on ratio, scale factor, similarity, and proportional reasoning.

- Sim Lemonade Millionaire (http://www.freeonlinegames.com/game/sim-lemonade-millionaire) engages students in a simulation game in which they run a lemonade stand. With an initial $100 to spend, players choose an advertising package based on percentage returns and a location for the stand based on population and mobility rate. Players purchase stock for the lemonade and create a recipe, determined by ratios of lemons, ice, sugar, and water. The free simulation includes upgrade opportunities, and players may save their games to continue at another time, allowing for long-term play.
- Ratio Rumble (http://mathsnacks.com/ratiorumble_game_en.html) is a free online game with a video game-like appeal. Players are given, on a game board, a ratio recipe in both symbolic and diagram form, and they must make an equivalent ratio by choosing colored potion bottles on the board. There are fourteen levels to the game, beginning with two-part ratios, moving to three-part ratios, and then to ratios with decimals. Students may save their games so they can start at an appropriate level during the next session. An accompanying teacher's guide and video provides teachers with an introductory lesson plan.
- Scale City: The Road to Proportional Reasoning (https://www.ket.org/scalecity), a free online game, engages students in an environment in which they go on a road trip throughout the city, learning about how scale factor is used in real-world situations. Each stop on the trip involves students in watching a video and interacting with an online project. Students are encouraged to work together as they participate in the interactive activity. For example, as participants in the World Chicken Festival, players scale up recipes; at the Kentucky Horse Park, visitors learn about direct and indirect proportions using distance, rate, and time. The Scale City Quiz uses a vending machine game as the tool to test students' knowledge after completing the road trip. The teacher's guide provides a rich set of resources, including lesson plans and directions for the simulations and games.

Expressions and Equations

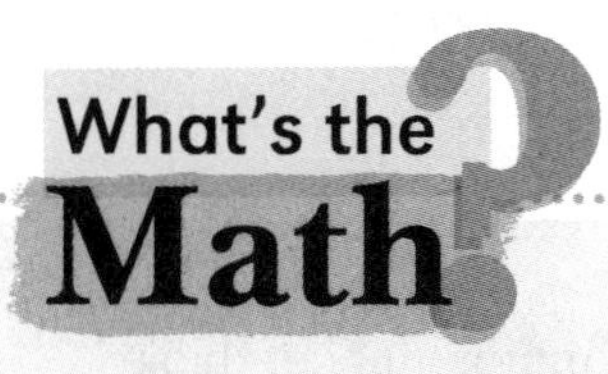

The study of expressions and equations helps students to transition from their work with number and operations in elementary school to the level of conceptual understanding needed for algebra. Previously, students have applied the properties of operations to rational numbers; now, they apply these ideas to variables. Success in this area is essential for later achievement in high school mathematics and in solving real-world problems. The National Research Council defines two key features of algebra as "(a) a systematic way of expressing generality and abstraction including algebra as generalized arithmetic and (b) a guided transformation of symbols such as we do when we solve equations by collecting like terms and using inverse operations" (NRC 2001, 256).

As students work with expressions and equations, there is ample opportunity for them to reason quantitatively and abstractly as well as to construct and understand viable arguments. Attention to justification helps students to connect prior knowledge about numbers to that of variables. According to Michael Cioe and his coauthors (2015), justification goes well beyond students showing their work, as it involves *why,* not just *how.* Playing and solving these puzzles and games in teams requires students to justify their thinking and to analyze the thinking of others.

In sixth grade, students read, write, and evaluate expressions, and translate an expression such as *6 less than x* to $x - 6$. Students use the properties of operations to recognize and generate equivalent expressions and to solve one-step equations and inequalities. They apply these ideas to the solution of real-world problems. In seventh grade, students extend these concepts and skills to the inclusion of rational numbers and the solution of multistep, real-world problems with positive and negative rational numbers. Students at this grade level also use mental computation and estimation strategies to assess the reasonableness of answers. In eighth grade, students solve linear equations efficiently and solve systems of linear equations with two variables. Students also make connections with other units of study as they use equations to express proportional relationships and apply their knowledge of functions in the interpretation of linear relationships (Gartland 2014).

Find It Together

Why This Game or Puzzle?

Communicating mathematical ideas and sharing mathematical thinking are essential habits of mind. For this puzzle, team members must share what they know in order to have all of the necessary data to find the solution. The need to collect data is relevant to almost all real-world problem-solving situations, as is the need to cooperate.

Here, a logic number puzzle is solved cooperatively, based on a model for cooperative problem solving suggested by Tim Erickson (1989). Each member of a three-person, cooperative team is given two clues about a real-world situation. The puzzle solvers must decide how to share and organize the information so they can create an equation and find the solution. Two sets of puzzle clues are provided, with Puzzle A including a fractional coefficient within the equation and thus is most appropriate for seventh and eighth graders, while Puzzle B is best for sixth graders.

***Find It Together* Puzzle Cards A**

On Wednesday the 15th of June, Mayara bought 6 pencils.	On Sunday the 19th of June, she had 20 pencils left.
On Saturday the 18th of June, she found 5 more pencils in her drawer.	On Friday the 17th of June, she gave half of her pencils to her soccer teammates.
On Thursday the 16th of June, she bought three times the amount she had on Tuesday the 14th of June.	How many pencils did Mayara have on Tuesday the 14th of June? Record the equation determined by the group.

Math Focus

- Writing expressions and equations to solve real-world problems
- Solving one-variable multistep equations

Materials Needed

- 1 set of *Find It Together* Puzzle Cards A or B per team of three students (page A-38 or A-39)
- Optional: 1 *Find It Together* Directions per team (page A-40)

Directions

Goal: Use the clues to find the mystery equation and solution.

- Form a team of three puzzle solvers.
- Place the clues facedown. Each solver randomly takes two of the clues.
- Solvers may read their clues to the others, but may not show the clues.
- Work together to figure out the equation and then the solution being described by the clues. Read the clues as many times as necessary, and talk about what you know.
- You can write or draw to help you understand the information in the clues, but you can't record the clues.
- When you think that you have the equation and the solution, read the clues again to check.

How It Looks in the Classroom

A sixth-grade teacher begins his class by introducing a simpler version of the puzzle that the students will be solving in their groups. He has created a puzzle with four clues, as shown in Figure 5.1. As students enter the classroom, they notice four cards placed face-down and taped to the whiteboard. He identifies four volunteers who are interested in solving a puzzle and asks them to sit at the table in the front of the room with paper and pencil. He tells the remainder of the students to sit at their desks as careful observers. The class quickly recognizes that they will be learning their new puzzle today by using a "fish-bowl" format, where the outer group observes the inner group as they solve. The teacher tells the outer group to identify the mathematics needed to solve the puzzle and to notice how the students work together.

Bertram paid $27 on Wednesday to rent a bike.	In addition to the rate per day, there is a fee of $3 for each 30 min (or part thereof) the bike is returned late.
Write and solve an equation to determine the number of minutes that Bertram was late.	Bailey's Bikes rents bikes for $18 a day on weekdays and $22 a day on weekends.

Figure 5.1 Mini-*Find It Together* puzzle

Each student at the front table is asked to choose a card from the board and to not yet share what is written on the card. Students are told that they may read the card aloud when it is their turn, but they may not show the card to the others. The teacher knows that Kylie will have a difficult time not sharing her card right away, so although she is asked to pick her card last, she is given the opportunity to read her card first. She reads, "In addition to the rate per day, there is a fee of three dollars for each thirty minutes (or part thereof) the bike is returned late," and Pedro immediately says, "I hope there's another card that says something about the rate per day."

Amber pipes up and says that she has a card about the rate per day, and then reads her card aloud. The group is more focused on the task as they hear Pedro read his clue about writing an equation in order to find out how many minutes late Bertram was in returning his bike.

As the inner group completes the puzzle, the teacher notices that the students in the outer circle are being careful observers and appear to be as engaged as the puzzle solvers. The observers later comment that they thought Amber did a good job of saying her clue just when it was needed.

The teacher asks students if they have questions about the mathematics of the puzzle or if they need further clarification about how they are to work together. When everyone is ready, the students form their groups to solve Puzzle A, and many of them are already talking about how they can help each other to not feel tempted to look at each other's clues.

Tips from the Classroom

- Most students are likely to want pencil and paper for taking notes about the clues and possible equations.
- Through field-testing we found that some students wanted to show each other the cards or write down the clues word-for-word, despite the requirement to only read the clues aloud. You may want to reinforce the goal of having students listen to one another carefully as they read the clues.
- We found that some groups, after solving the initial puzzles, wanted to create their own sets of clues to exchange with another group.

What to Look For

- Do students flexibly use diagrams, models, written words, and algebraic expressions? Figure 5.2 shows one group's use of words in combination with an algebraic expression to demonstrate its thinking about Puzzle A.

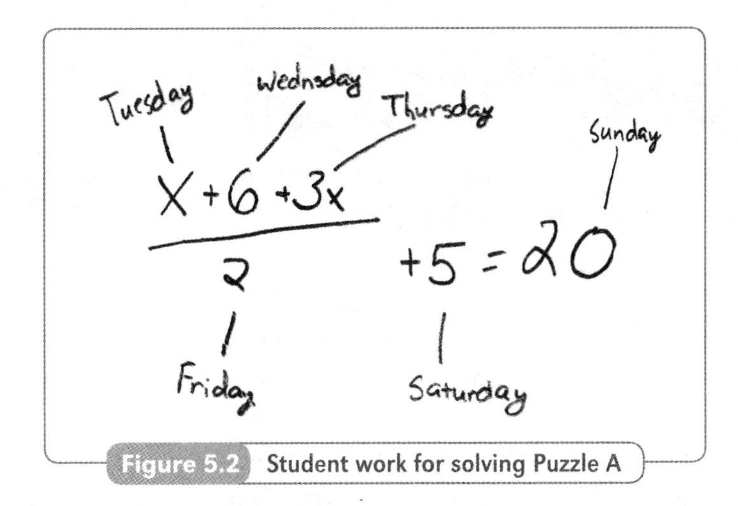

Figure 5.2 Student work for solving Puzzle A

- What strategies do students use for organizing the information they are given?
- How do the students work together? Do they make sure everyone participates beyond just reading the clues aloud? Do roles such as facilitator and note taker evolve as the students work on the puzzle?

Variations

- Have solvers show their clues one at a time, without talking or gesturing. After all the clues have been shown, each solver may rearrange the order of one of the clues. Through eye contact, rather than talking, solvers communicate that they think they have found the equation and a solution. Solvers may then talk to check the equation and solution.
- Solvers may choose several but not all of the clues and try to determine the missing information needed to solve the puzzle. As another clue is revealed, they determine if that information was part of their list of missing information.
- If you, or your students, choose to make your own *Find It Together* clue cards, you could increase or decrease the number of cards and the complexity level of the equation needed to find the solution.

Exit Card Choices

- You are given information in a puzzle about the number of hours that you babysat and the total amount of money that you made. What could you be solving for? Explain.
- Write about how your team decided to work together to find the equation and the solution to your puzzle.

Extension

Have students contribute a phrase or a picture to a poster titled *Cooperative Puzzle Solving*.

Equivalent Expressions

Why This Game or Puzzle?

This game requires students to match expressions that are equivalent. Players might do so by recognizing one expression as a simpler form of another, or by applying their understanding of the properties of arithmetic to algebraic expressions. When students examine the structure of expressions before immediately computing, they may discover an efficient way to identify those that are equivalent. The ability to look for and make use of structure is an important aspect of mathematics proficiency (National Governors Association Center for Best Practices (NGA) and Council of Chief State School Officers (CCSSO), 2010).

***Equivalent Expressions* Cards**

$-(x - 12)$	$12 - x$
$2x - 2 + 5$	$3 + 2x$
$2x - (-2) + 24$	$2(x + 13)$
$7x + 5x - 2$	$-2 + x(7 + 5)$
$2x - 7$	$-3 - 4 - (-2x)$
$-3(x - 7)$	$30 - 3(3 + x)$
$20 - 3x - 1 + 2$	$-3x + 21$
$12x + 15y$	$3(4x + 5y)$
$15y + 12x$	$-3(-4x + (-5y))$

After finding a match, students are asked to write the two expressions on a recording sheet. Recording pairs of expressions provides students with a list from which underlying structures, such as the properties of arithmetic, can be discussed. The game cards include a variety of difficulty levels and, as explained in the "Variations" section, the game can be played with a subset of the cards.

Math Focus

- Applying the properties of arithmetic to algebraic expressions
- Simplifying expressions
- Finding equivalent expressions

Materials Needed

- 1 deck of *Equivalent Expressions* Cards per group (pages A-41 to A-43)
- 1 *Equivalent Expressions* Recording Sheet per team (page A-44)
- Optional: 1 *Equivalent Expressions* Directions per group (page A-45)

Directions

Goal: Collect more pairs of equivalent cards (packs).

- Shuffle the cards. Deal each team four cards faceup for all to see. Put the other cards facedown in a pile.

- On each turn you can do one of three things:
  1. Find two of your cards that have equivalent expressions. Set this pair beside you. Replace them with two cards from the top of the pile.
  2. Trade one of your cards with one of the other team's cards when you are able to make a pair by doing so. Set this pack beside you. Each team replaces its card with a card from the top of the pile.
  3. Draw a card from the top of the pile and add it to your cards.
- When a pair is made, both teams must agree that the expressions are equivalent and then one player records their thinking on the recording sheet.
- If no cards are left in the pile, you can still have a turn, but you don't replace any cards you play.
- The game ends when no team can make another pair.
- The team with more pairs wins.

How It Looks in the Classroom

One sixth-grade teacher writes the following list of expressions on the board and asks students to work with a partner to find two expressions that are equivalent.

$$x + 6$$
$$4 - 3(x + 6)$$
$$4 - 3x - 18$$
$$-3x - 14$$

Henry and Daniella work together and are overheard conferring with the other partners at their table about whether you should subtract $4 - 3$ first. When it is time to discuss as a whole group, Daniella reports that the third and fourth expressions are equivalent. The teacher asks Avry, who nodded his head when Daniella shared their results, to state why he agrees and he says, "Because the fourth expression shows combining the terms that are alike from the third one."

Hideo reports that he and his partner think that the first and second expressions are equivalent. Ciarra says, "I'm confused, because I think that the first and second expressions are equal and the second and third are, but I don't think that the first and third are, so how can they all be equal? Something's wrong."

The teacher asks the entire class to consider what Ciarra has just said, and then asks Ciarra to repeat her thinking so that students can write it down if they want to do so. After a few minutes of independent thinking, the teacher tells the students that they can confer again with their partners if they wish, and then asks for volunteers to describe their thinking. Marko says, "We're not sure if this is right or not, but Alina and I think that it is, so we'll try. We think that the first and second expressions aren't equal because you don't subtract 3 from 4 first. The 3 needs to be multiplied by what's inside the parentheses first. That should fix Ciarra's confusion about why they can't all be equal. And, that means the last three expressions are all equal!"

The teacher compliments the students on their willingness to help each other clarify their thinking, especially when it's faulty, and tells them that this kind of discussion will help them to play *Equivalent Expressions*. She has the students at one table model how to play the game while others observe, and then hands out the materials for students to play in groups of four, with partners.

Tips from the Classroom

- At first we were uncertain as to whether students should pause play to record their matching expressions. Through field-testing we found that it did slow down the game, but that the slower pace encouraged attention to whether or not the expressions were equivalent.
- Games may require more than one copy of the recording sheet, so you may want to make double-sided copies.
- Encourage students who are unsure whether two expressions are equivalent to replace the variable with a one-digit number and compute an answer. This approach is an example of simplifying a problem, an important problem-solving strategy.
- Students may want to use scrap paper as they play.

What to Look For

- What strategies do students use to find equivalent expressions?
- Do students recognize examples of the commutative and distributive properties?
- What language do students use when deciding whether or not two expressions are equivalent?
- Do players consider what cards their opponents have when they decide which card to give away?

Variations

- For sixth-grade students, use only the cards on pages A-41 and A-42.
- For upper grade levels, you could use only the cards on pages A-42 and A-43. You can also choose to create your own subset. Just remember to keep a matching card for each card you include in the set.
- In our field-testing, some students suggested that if a team chose a pack with two cards that were not equivalent, the team would lose its turn.

Exit Card Choices

- How do you know that $3(2x - (-5)) = 15 + 6x$?
- What two expressions can you write that are equivalent to $4 - 5(x + 6)$?

Extension

Post several expressions throughout the room and have a scavenger hunt. Teams can use the *Equivalent Expressions* Recording Sheet (page A-44) to record those they find that are equivalent.

Solve It Game Board

Equations:	Solutions:
____x + ____ = ____x - ____	x = ____
____ - ____x = ____	x = ____
____(____x - ____) = ____	x = ____
____(____ + ____x) = ____x	x = ____
____x + ____ = ____ - ____x	x = ____

Discard ____ Sum of the values for x = ____

Why This Game or Puzzle?

The *Solve It* game board provides spaces to write nineteen numbers (and one discarded number) which, when written, create five expressions. The game leader shuffles the game cards made from a red suit (negative numbers) and a black suit (positive numbers) of playing cards, 1 (ace) through 10. The top card is then turned over and each team writes the number in one of the open spaces on its game board. Play continues until all numbers are placed. (Note that one number is written in a discard space.) Players then solve for x and find the sum of their x-values. The team with the lesser (or greater) total value wins.

Students are often surprised at what a difference the placement of the numbers makes. Opportunities to consider the impact of choosing whether a number should be a coefficient or a constant begin to build students' sense of algebraic structure. As students play, they gain a better understanding of the relationships among the numbers, symbols, and structure within an algebraic expression. After a few games, players often make conjectures, such as *Sometimes choosing a greater number for the coefficient with a negative value will result in a greater value for* x or *We have the chance to make this an equation with infinite solutions.* Jarmila Novotná and Maureen Hoch (2008) consider structure sense to be an extension of symbol sense, which they view as an extension of number sense, and they believe it is key to later success in high school algebra and college mathematics.

Math Focus

- Using properties of operations to solve linear equations
- Solving one-variable equations employing multistep processes

Materials Needed

- 1 *Solve It* Game Board per team (page A-46)
- 2 suits (one red and one black) from one deck of standard playing cards, face cards and jokers removed, per group (black cards represent positive numbers, red cards represent negative numbers, ace = 1)
- Optional: 1 *Solve It* Directions per group (page A-47)

Directions

Goal: To score either the lesser (or greater) number of points as a result of solving equations.

- Decide if the lesser or greater number of points wins.
- Shuffle the cards and place them facedown. Choose a player to also be the game leader.
- The game leader turns over one card and announces the number. Each team writes the number in one of the spaces on the game board. A discard box is provided for one of the numbers.
- Both teams must record the number before the next number is announced and, once written, its placement cannot be changed.
- The game leader continues turning over cards until twenty numbers have been recorded.
- Both teams then solve the equations that were created by recording numbers in the spaces.
- Each team finds the total of its x-values. If an equation results in a value for x that is not possible, the team adds 0 points to the total. For each equation that results in a value for x for which the solution is all real numbers, the team scores -5 (or 5) points. The team with the lesser (or greater) sum is the winner.

How It Looks in the Classroom

This eighth-grade teacher knew that her students were, after a busy week of standardized testing, in need of some action. She decided to create a new type of game rollout, her usual term for introducing a new game. After her students complete a warm-up on solving a multistep equation, she asks them if they have ever seen the TV game show, *The Price Is Right*. Many students raise their hands and say that they like to play along as the contestants try to find correct prices. One student says that she likes to play the game where they have to make equations to find the right price.

The teacher tells the students that they will be playing a game similar to one played on *The Price Is Right*. She first unveils an exciting prize (one extra credit point) that one lucky contestant will win if he or she chooses the right number, and the students excitedly raise their hands to be chosen as the contestant. Alejandro is chosen and moves quickly to the "stage," as he has seen the contestants do on TV.

The teacher unveils a game board as shown in Figure 5.3 and tells Alejandro that he must choose the correct number to place in the empty spot in order to create an equation that results in the least value for *x* when solved. The "audience" is calling out possibilities, as they do on the game show, and Alejandro focuses on which missing coefficient will work best with the numbers that are already in the equation. The teacher tells him that he can ask one audience member for advice and he calls on Maggie. She comes to the stage and the class overhears them talking about how using the least number will result in a final value of *x* that is greater.

Numbers to choose from: 2 3 5 7

$-\_\_\_\_x + 4 = 5x - 2$

Figure 5.3 Game board used for introducing *Solve It*

Several other students are heard questioning each other as to why this is true. Brendan and Linese discuss, at their table, how the negative sign in front of the missing coefficient means that they will have to add the inverse, a positive value, to both sides of the equation, creating a greater coefficient on the right and resulting in a greater denominator. Alejandro decides to choose the number 7 and the audience is asked to check if this is the correct number. After agreeing as a class that it is, Alejandro happily leaves the stage with his prize.

The teacher tells the students that they are now all going to be contestants by playing *Solve It* in teams, and she explains the similarities and differences between the game they just played as a class and the game they will play in teams. The students clearly enjoyed the way in which the game was introduced and ask if they could make up more games like this themselves.

Tips from the Classroom

- Versions of *The Price Is Right* are available online. Prior to this game introduction, you may want to show your students a brief segment of one episode so that all become familiar with the nature of this game show.
- Students may wish to play several rounds of this game within one sitting, as each time they play they learn to make better choices as to where to write the numbers. Consider laminating the game boards or copying the board on both sides of the paper.
- Some teams may take a long time to decide where to write a number. Using a timer, with a predetermined amount of time, can be helpful in keeping the game moving.
- Consider students' pace when determining teams. A quick or impulsive thinker paired with a slower or methodical thinker may or may not make a good partnership.

- You may wish to have students use a pen to record their numbers so that it is less likely they will change an earlier number choice.

What to Look For

- How are students thinking about the signs of numbers when they decide where to place them?
- What conjectures or generalizations do students make about the placement of numbers; for instance, do they notice when it is best to have a lesser or greater number as a coefficient?
- What evidence do you observe of partial understandings or common errors? For example, do students seem confused when a negative number follows a subtraction sign?

Variations

- If you would like to play this game without negative numbers, just use the black cards twice.
- You could choose to include fractions as coefficients, which would necessitate adding more cards.

Exit Card Choices

- You must fill in the following equation using the numbers 3, 5, 2, and 1. Find the greatest possible value for x and then the least possible value for x.

  $\_\_\_\_ - \_\_\_\_x = \_\_\_\_x + \_\_\_\_$
- A student solved the first step of the equation $5 - 3(x + 2) = 10$ as $2(x + 2) = 10$. Is this correct thinking? Why or why not?

Extension

- Once teams have placed their numbers, give the homework assignment of rearranging them to improve their scores.

Why This Game or Puzzle?

Students, even those who are successful at manipulating algebraic expressions, often find it challenging to translate between words and algebraic notations. Their ability to recognize the same structural relationship shown in different representations is a significant indicator of conceptual understanding (Panasuk, 2011). *Express It!* gives players many opportunities to talk with their peers as they wrestle with the ideas involved in such translations.

The game begins with players placing a set of cards—each showing a number, variable, or symbol—faceup in an array between their two teams. Then players shuffle a deck of written expressions, place it facedown, and turn over the top card. Each team attempts to represent that expression using the numbers and variables in the array, along with its team's symbol cards. The first team to think it can do so arranges the cards and, if the opponents agree, keeps the number cards used in a set. Play continues until no more sets can be made. The written expressions in Card Set A are relatively short and abstract, such as *The difference of three times a number and ten*, while those in Card Set B contain more words as they represent real-world contexts.

***Express It!* Written Expressions Card Set A**

1 Thirty-two divided by the quantity two times a number plus four	2 The difference of three times a number and ten
3 The square of a number increased by six	4 Six less than the product of two and a number
5 Three times the quantity of three plus a number	6 The sum of twice a number and five times the same number
7 The quotient of seven squared divided by a number	8 The quantity of eight times a number plus two, cubed
9 Four times a number decreased by eleven	10 Eight divided by four times a number

Math Focus

- Using variables to represent quantities in a written expression
- Using variables to represent quantities in a real-world situation
- Reading and creating algebraic expressions

Materials Needed

- 1 *Express It!* Written Expressions Card Set A or B per group (page A-48 or A-49)
- 1 *Express It!* Numbers and Variables Cards A or B per group (page A-50 or A-51)
- 1 *Express It!* Symbols Cards A or B per team (page A-50 or A-51)
- 1 *Express It!* Recording Sheet per team (page A-52)
- Optional: 1 *Express It!* Directions per group (page A-53)

Directions

Goal: Given target expressions (in words), create as many expressions (in algebraic form) as possible before all of the cards are used.

- The Written Expressions Cards are shuffled and placed facedown in a deck. The Numbers and Variables Cards are arranged on the playing area faceup in an array. Each team is given a set of Symbols Cards.
- Teams alternate turning over a card from the deck.
- Each team tries to use the number and variable cards as well as its own symbol cards to represent the written expression, without moving the cards. When a team believes it has done so, its players call "Express it!"

- At that point, the team arranges the number and variable cards, and its symbol cards, to make the expression. The other team checks for accuracy.
- If the team is correct, it keeps the number and variable cards it used as a set, retrieves its symbol cards for reuse, and records the expression on its recording sheet. If it is not correct, the team must put the number and variable cards back into the array, retrieve its symbol cards for reuse, and then both teams keep looking for a correct expression.
- After a correct set is made, a team turns over a new Written Expression Card and play continues.
- When there are no more number and variable cards remaining or when no more sets can be made, the team with the greater number of card sets wins the game.

How It Looks in the Classroom

This seventh-grade teacher knows that his students enjoy working together to solve problems by talking about what they do understand and what they are uncertain about. To introduce the game they will soon play in teams, he puts an algebraic expression, written in words, under the document camera, and tells students that they must find the correct cards, posted on the board (Figure 5.4), to write the expression in algebraic form.

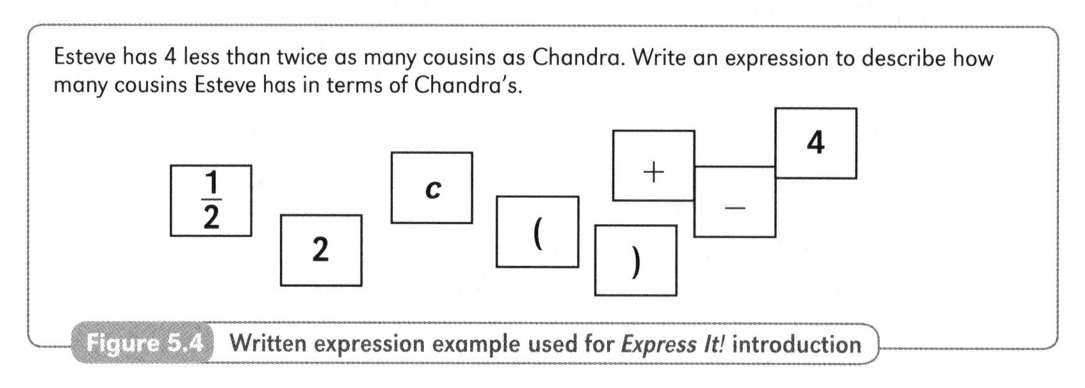

Figure 5.4 Written expression example used for *Express It!* introduction

Students work in groups of three, at their tables, to discuss their interpretation of the expression. After an appropriate amount of time, the teacher asks two groups to post their completed expressions on the board and the class is asked to observe the similarities and differences between the two expressions. The group members who posted $\frac{1}{2}c - 4$ are eager to describe why they think their answer is correct; however, the teacher asks another group to comment on this response. One member of that group volunteers that the *–4* is correct, but the $\frac{1}{2}c$ would mean that Esteve has only half the number of cousins that Chandra does, not twice as many. Now that the first group has heard the alternative stated aloud, two of the students in that group agree that the correct coefficient for c would be 2, not $\frac{1}{2}$.

Another student in the group holds out and says that he needs to try the expression with real numbers in order to see if it works. The whole class agrees to try this method, using 5 cousins as a possible amount for Chandra. Several groups then report that they agree the correct expression is $2c - 4$, and one group reports that they used a bar model to further prove that the expression works.

This teacher is pleased that the students persevered to solve the problem and worked collaboratively to consider various methods for finding the correct expression. After listening to the directions for how to play the game in their groups, some of the students are heard reminding each other how they could use a bar model to represent the expressions and others take some color cubes off the shelf in case they need them.

Tips from the Classroom

- Some students may need to make models or diagrams prior to creating the expressions algebraically. Encourage students to take the time to make diagrams, if this helps them to find the algebraic expressions more accurately, and then provide more experiences with transitioning from the written expressions to the algebraic expressions directly.
- Sometimes teams may allow their opposing team to choose the expression without checking it for accuracy. Encourage students to check each other's expressions as a way to further their conceptual understanding.
- You may want to copy the symbols on a different color of paper so that students can readily identify them for reuse.
- Some students would benefit from having the cards on stock paper for easier movement.

What to Look For

- What level of confidence do students exhibit as they create the expressions? Are they making a number of expressions before they choose the correct one? Are their choices reasonable?
- If you find that one team is working much faster than another at creating the expressions, you may wish to set a timer and allow both teams to have a certain amount of time prior to calling "Express it!"
- How do students talk about the mathematics they are using as they find the expressions? Do they use language that shows they recognize various properties?

Variations

- You can vary this game by changing the level of complexity of the expressions, having students find expressions involving negative exponents or fractional coefficients.
- You may choose to keep score by counting the number of cards rather than the number of sets.

- You may want to have students play the game in reverse, with new cards, by having the written expressions given in parts on separate cards and the algebraic expressions provided on the target cards.

Exit Card Choices

- Use words to represent the expression $3x - 2$. Find two different ways.
- Maribel made $25 raking leaves yesterday. She raked for 3 hours and got a $4 tip. Write an expression to represent the amount Maribel earned each hour she raked leaves.
- A student created the algebraic expression $2 - 6x^2$ to represent the written expression *The quantity 2 minus 6x, squared*. Another student said that wasn't right, but didn't explain her thinking. What do you think, and why?

Extension

Given the topic *Ways to help me translate from words to into algebraic expressions*, have students write their top three ideas in their math journals, along with specific examples. Provide time for students to share their thinking in pairs or small groups.

Linear Systems Bingo

Why This Game or Puzzle?

According to Marian Small, "We have become a highly visual culture in which pictorial images have begun to supplant the printed word as our preferred way of navigating the world," (2013, 1). In this game, visual images on domino-like cards represent a system of two equations. We found that the visual models were more motivating to students than equations and that the images allowed students to "see" relationships they might not otherwise recognize.

To play, each team writes *FREE* in one of the empty spaces on its *Linear Systems Bingo* game board and then randomly writes the fifteen given answers in the remaining spaces. The fifteen game cards are placed faceup between the teams in a three-by-five array. On its turn, a team chooses one of the cards, each of which shows a visual model of a linear system with two variables. Both teams then work independently to find the solution to the system represented on the card, or to find that there is no solution or infinitely many solutions. When everyone agrees on the answer and discusses the solution strategies that were used, they mark the answer on their game boards. The first team to have marked four solutions correctly in a row, column, or diagonal, wins.

Math Focus

- Representing a visual model as a linear system with two variables
- Solving systems of linear equations
- Explaining why graphing, substitution, or elimination was used to find a solution

Linear Systems Bingo Game Board

Make your choice of where to place a FREE space.
Then randomly choose the cells in which to write the following answers: (1, 3); (12, 3.5); (2, 4); (2, 9); (3, 2); (6, 4); (9, 22); (2, 3); (5, 7); (1, 1); (12, 3); infinite solutions; infinite solutions; no solution; and no solution.

Materials Needed

- 1 *Linear Systems Bingo* Game Board per team (page A-54)
- 1 deck of *Linear System Bingo* Cards per group (pages A-55 or A-56)
- Optional: Graph paper, as needed
- Optional: 1 *Linear Systems Bingo* Directions per group (page A-57)

Directions

Goal: Be the first to get four answers, or three answers and a FREE space, in a row, column, or diagonal.

- Each team privately completes its copy of the game board by writing *FREE* in one space and then randomly writing the solutions provided in the blank boxes.
- Shuffle the *Linear System Bingo* cards and place them faceup between the teams in a three-by-five array.
- One team chooses a card. Note that the mass of the same shape is equivalent on one card, though not necessarily equivalent to the mass of the same shape on another card.
- Both teams find the solution and discuss the answer and their solution strategies. Note that, though either shape could represent either coordinate, in this game the cylinders will always represent the *x*-coordinate and the cubes will always represent the *y*-coordinate.
- When all players agree, the teams cross out that answer on their game boards and the card is turned facedown.
- Teams alternate choosing a card. The first team to get four answers in a row, column, or diagonal says, "Bingo."
- If both teams agree that all of the answers crossed out are valid, the team that said "Bingo" wins.

How It Looks in the Classroom

To introduce *Linear Systems Bingo* in one eighth-grade classroom, the teacher projects cards A and B shown in Figure 5.5, which are similar to those used in the game. The teacher explains that on any one card, the mass of each cylinder is the same and the mass of each cube is the same. She further notes that cylinders and cubes on other cards may or may not have that same mass. She asks students to determine whether the two cards represent a system with one solution, with no solution, or with infinitely many solutions. She observes that some students begin to immediately write equations, while others are involved in discussions with partners, which she thinks suggests that an interesting conversation will ensue.

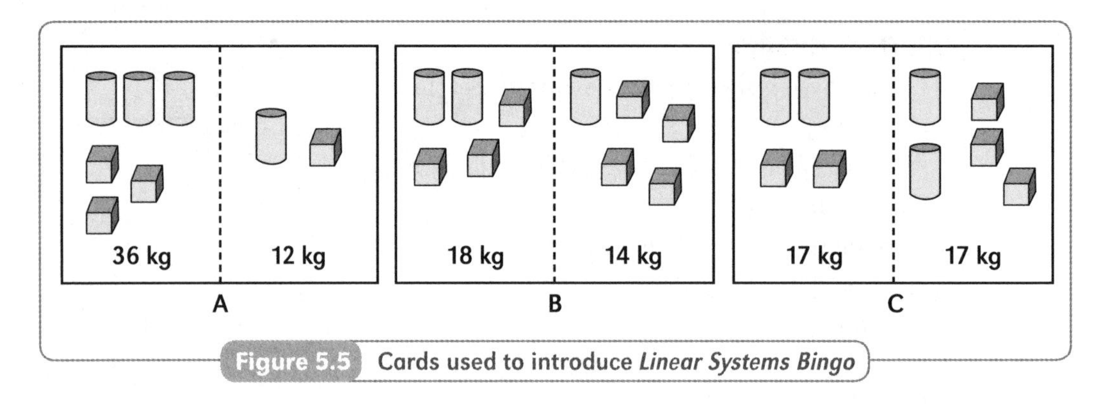

Figure 5.5 Cards used to introduce *Linear Systems Bingo*

The teacher regains their attention and asks the students to share what they discovered about card A. Patrick says that it has infinitely many solutions and the teacher asks how he knows. He explains that he wrote the two equations, multiplied each term in the second equation by three, and found out that they were the same equation. He concluded that any x and any y would work in the system. Many heads nod in agreement, presumably because several students drew the same conclusion and followed the same strategy.

The teacher asks if anyone solved it differently. Jexx explains that he and his partner agree, but solved it a different way. He reports, "We made a graph for each image, remembering that the mass of the cylinders were x and the cubes, y, and found that the lines were drawn on top of each other." Dionice explains how she and her partner decided just by looking, as the number of cylinders, cubes, and kilograms are all in a 3:1 ratio. Monroe adds, "We were thinking the same thing, almost, except in reverse, because we saw that everything was one-third on the right as compared to the left." Patrick sighs and says, "I wish I saw that."

They review card B as a class and agree that the solution is (6, 2). Then the teacher reveals card C and tells students to look at it critically, before writing anything, and then to turn and talk to their partners about what they think. Though some students eventually write equations, more students recognize that it is not possible to have everything the same except for the number of cubes, so there could not be a solution. The teacher then explains the game, distributes materials, and instructs students to begin the game.

Tips from the Classroom

- Make sure that teams discuss and agree upon their solutions before marking answers on their game boards.
- Students do not always choose the most efficient solution strategies. We found that posing questions or making comments, such as *Hmmm, I wonder if there is a more efficient way to solve this one*, motivated students to stop and consider their approach.
- When students find a card that has no or infinite solutions, encourage them to discuss why they think that is the case.
- You may wish students to use a timer so that teams do not spend too much time choosing a card.

What to Look For

- Do teams' solution strategies vary, depending on the specific combination of equations? Can they explain why they chose a particular technique?
- When teams are choosing a card, do they talk about which combinations might have one solution as well as discuss those with no or infinite solutions?
- What do teams do when they disagree on an answer? For example, do they merely accept the one that matches an answer on the game board or do they take the time to explain why that one is correct?
- What vocabulary do students use as they explain their solution strategies?

Variations

- You could allow play to be more random, with the cards placed facedown.
- You can change the game to a puzzle by giving solvers the sheets of uncut cards and a copy of the list of solutions. In partners, solvers identify a correct solution for each card.
- Replace the visual models with word problems about total cost (or a similar topic), using the same equations as shown visually.

Exit Card Choices

- Marcus was playing this game with a different set of cards. He chose the card shown here and said, "This is going to be easy." What, in this image, might have caused Marcus to make that comment?
- Given the system $4x + 8y = 18$ and $4x + 6y = 14$, what solution strategy would you use? Why?
- How did your partner help you to think about a problem in a different way?

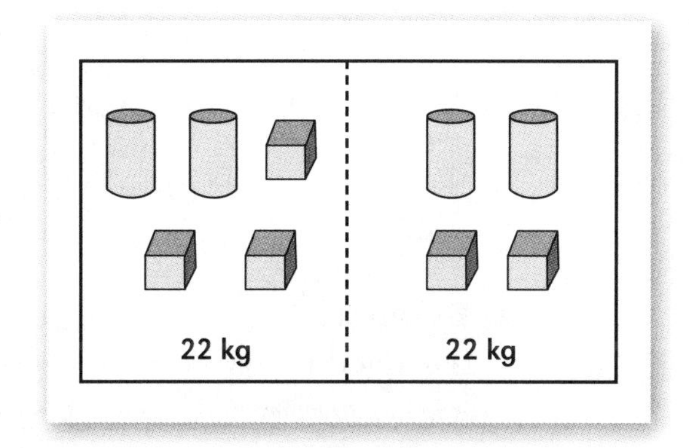

A student response to the third exit card is shown in Figure 5.6. Note the student's description of how her partner explained why there was no solution. Writing about her partner's thinking helped this student solidify these ideas and reminded her of how much she can learn from a classmate.

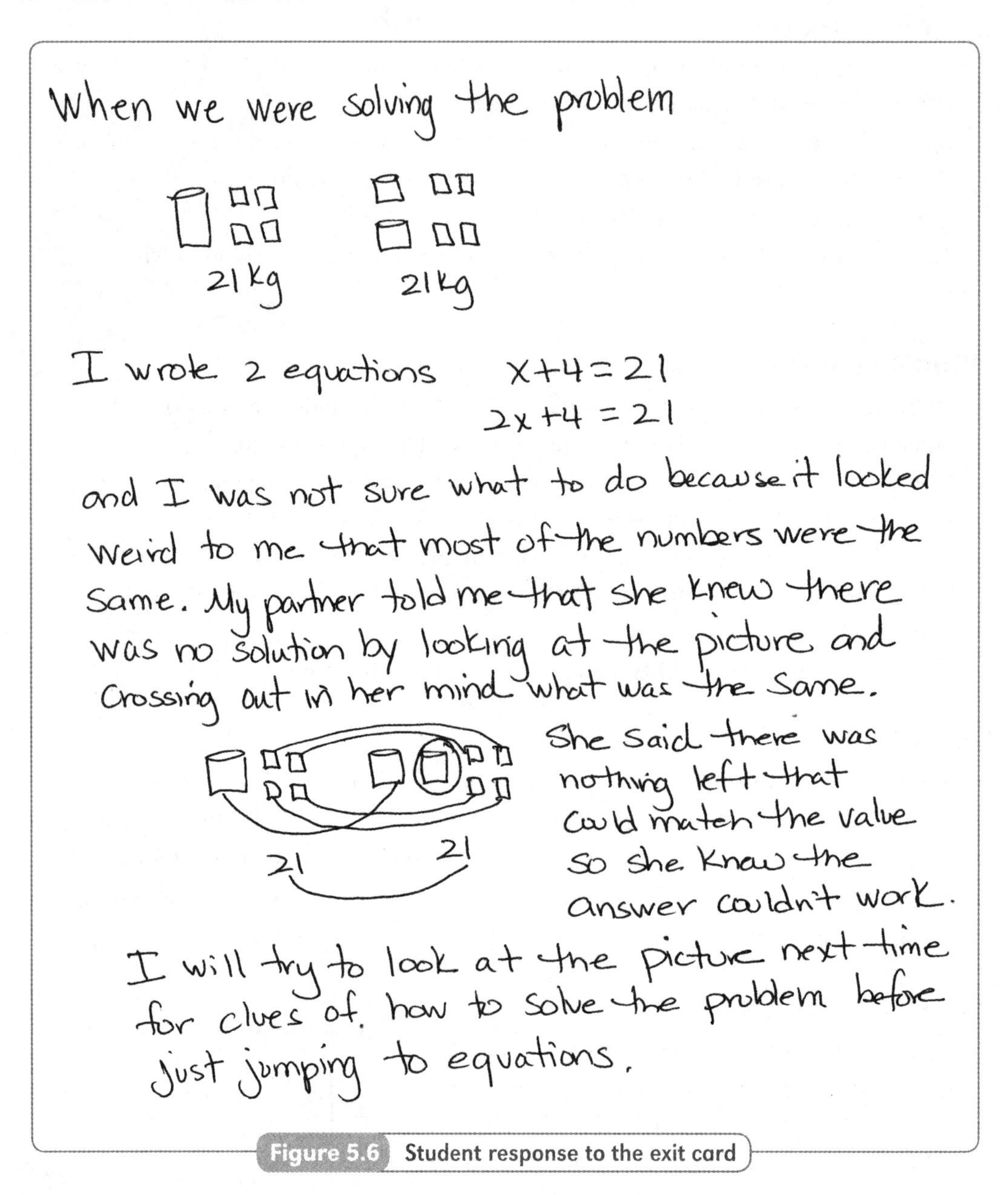

Figure 5.6 Student response to the exit card

Extension

Ask students to write in their math journals in response to the question: *If you wrote an equation to represent the image on each side of a game card, how would you describe the meaning of the slope and* y*-intercept for the graph of each equation?*

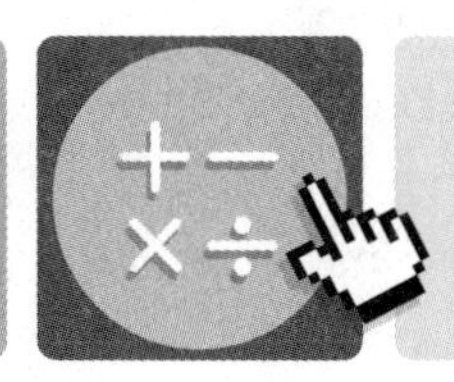

Online Games and Apps

Computer images are often useful aids as students transition from the use of manipulatives to that of pictorial models in learning algebraic concepts involving expressions and equations. While some online games and puzzles involve students in practice with combining like terms and solving equations through Jeopardy-type and matching games, others engage students in manipulating visual images and using visual-spatial strategies to learn algebraic concepts. Some online games and apps also provide students with opportunities to experiment, while getting feedback, as they discover and deepen their understanding of algebraic structures, and those games and apps with multiple levels of challenge can support individual learning needs.

- Dragon Box Algebra (5+ and 12+), a fee-based app (www.itunes.com), engages students in learning algebraic rules through experimentation and discovery by initially using animal-faced cards and then transitioning to numbers and variables. Students are provided with several cards and a box, and by following certain rules, they must eliminate the unnecessary cards in order to, essentially, isolate the variable. Each of ten chapters reveals a progression of icons, eventually replaced by numbers and variables, so that students recognize their work toward a solution as solving an equation. The more efficiently the player manipulates the icons, the greater the number of points obtained in the game. Teachers may access worksheets and related resources to enhance students' learning while playing this game.

- Played as a game or as individual puzzles, Lure of the Labyrinth (free from www.thinkport.org) engages middle school, pre-algebra students in a world in which puzzles are solved by maneuvering through a labyrinth, with the goal of finding a lost pet. The mathematics topics involved in solving the puzzles focus on a variety of algebraic reasoning concepts. There are nine different puzzles, with three levels within each of the puzzles supporting many different levels of student understanding. As an example, students solve The Employee Break Room puzzle by determining the monetary value of each of the "eyes" on the vending machine by creating expressions with known values. The goal is to purchase the described amount of food using the required amount of "eyes." In the Testing Lab puzzle, students are given a recipe to make by combining amounts of the given ingredients in measuring jars. Solvers may choose to create expressions to determine the most efficient method for making the recipe. Teachers are provided with sample lesson plans, which they may use to introduce the mathematics involved in each of the puzzles, as well as recording sheets for students to document their puzzle solution strategies.

- Algebra Puzzles (free from www.mathplayground.com) involves students in finding the values of various objects found in either a 3-by-3 or a 3-by-4 grid. The numbers found on the right of and below the grid are the sums of the values of the objects. Solution strategies may involve using visual thinking to find the mathematical relationships among the objects or creating a set of simultaneous equations to solve algebraically. As students must assign a numeric value for three objects, feedback is provided as to whether one, two, or all three solutions are correct, giving puzzlers the opportunity to continue working toward an accurate solution.

Statistics and Probability

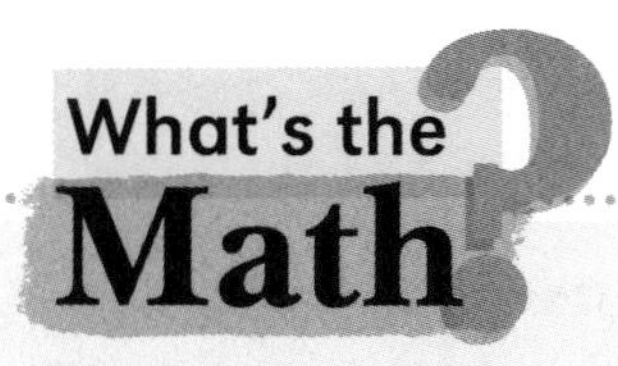

Every day we are inundated with a staggering amount of data and, according to the Bureau of Labor Statistics U.S. Department of Labor (2014–2015), employment of statisticians is likely to grow by twenty-seven percent in the next decade. Together, statistics and probability help us to make sense of our world and make predictions about the future. As Selmer and her colleagues suggest (2011), today's students must make life decisions based on their ability to draw conclusions from an abundance of data. We should not be surprised, therefore, to find that learning expectations for statistics and probability have increased and now incorporate many ideas that you may have learned in a college-level course, or that may be entirely new to you. The games and puzzles in this chapter provide opportunities for students to gain a deeper understanding of statistical measures and representations as well as to conduct experiments and consider theoretical probabilities.

At the middle school level, students collect and analyze data. Sixth-grade students explore ways to summarize quantitative data. They learn to describe data through their shape and center, while recognizing that finding means and medians are different ways to measure centers and may yield different results. Students also consider the spread of data and learn to display data in a number line plot, histogram, or box plot.

In seventh grade, students expand their ability to reason statistically by learning how sampling can be used to gain information about a population.

They learn to take random samples and to recognize the importance of representative samples if they wish to draw conclusions. Seventh graders also begin to explore probability, learning that all probabilities are represented as numbers between 0 and 1. They conduct experiments and develop probability models. They use simulations or make organized lists, tables, and tree diagrams to find the likelihood of more than one event occurring (compound probability).

In eighth grade, students consider relationships between two quantitative variables, known as bivariate data. Students create scatter plots and place a line through the data, looking for a best fit, to represent the association. For example, students might explore the relationship between age and the number of hours of sleep a night. Eighth graders also consider categorical data by making frequency tables to consider relationships among categories. Here, students might investigate whether there is evidence of an association between, for example, practicing yoga and eating healthful foods.

***Target Statistics* Recording Sheet**

Name(s): ______________________ Date: __________

Round 1:	Target	Five Data Points	Our Statistic	Points
Mean				
Median				
Range				

Round 2:	Target	Five Data Points	Our Statistic	Points
Mean				
Median				
Range				

Round 3:	Target	Five Data Points	Our Statistic	Points
Mean				
Median				
Range				

Why This Game or Puzzle?

There are a variety of ways to describe a set of data. Two of the most common ways are to use statistics to identify what is typical and to describe how the data vary. Too often, students are given a list of data and asked to merely apply procedural knowledge to it, such as finding the mean or median. As Stacey Reeder (2012) reminds us, we have to find engaging ways to explore mathematical ideas as they become more abstract.

Target Statistics focuses on the mean, median, and range of a set of data in a way that deepens conceptual understanding. Players use a standard deck of playing cards with the face cards removed and aces representing ones. "Targets" for the three statistics are determined by randomly choosing a card for each. Players then turn over fifteen cards, one at a time, each representing a data point. Teams privately decide the statistic under which to write each number, before the next card is shown. Decisions may not be changed. Next, teams determine the mean, median, and range for their data and receive a point for each of their statistics that are closer to the target. Throughout the game, players have opportunities to gain a better understanding of the impact one data point can have on different statistical measures.

Math Focus

- Finding the mean, median, and range of a given data set
- Creating data sets based on a given mean, median, or range

Materials Needed

- 1 standard deck of playing cards with face cards removed (ace = 1) per group
- 1 *Target Statistics* Recording Sheet per team (page A-58)
- Optional: 1 *Target Statistics* Directions per group (page A-59)

Directions

Goal: Choose cards to form data sets closest to three "target" statistics: the mean, median, and range.

- Shuffle the cards.
- Turn over the top three cards of the deck for all players to see. The first card identifies the target mean; the second card gives the target median; and the third, the target range. Each team writes these "target" statistics on its recording sheet.
- A player from one team turns over the next card of the deck and each team decides privately whether to record the number in the data set for the mean, median, or range. Once a number is written, it may not be changed.
- Once both teams have recorded the number, the other team turns over a card and again, teams decide where to write this number.
- Cards are turned over and the numbers are recorded until each statistic has exactly five data points.
- Each team determines the mean, median, and range for the data sets it created.
- If one team's statistic is closer to the target, it earns 1 point. If there is a tie, then both teams earn 1 point.
- The team with more points after three rounds wins, or there is a tie.

How It Looks in the Classroom

The students in one sixth-grade class have learned about the meanings of median, mean, and range and how each statistic is computed, but have not yet had an opportunity to manipulate data in order to create a statistic. The teacher introduces the game *Target Statistics* by letting students know that, not only is this their chance to practice finding the mean, median, and range, but more importantly, to learn how each of these statistics is affected by introducing another number to the data set.

The teacher begins by writing the number *6* on the board and telling students this number will be their target median. She then gives the students a set of ten data points as shown in Figure 6.1, and tells them to choose five data points with a median as close to 6 as possible.

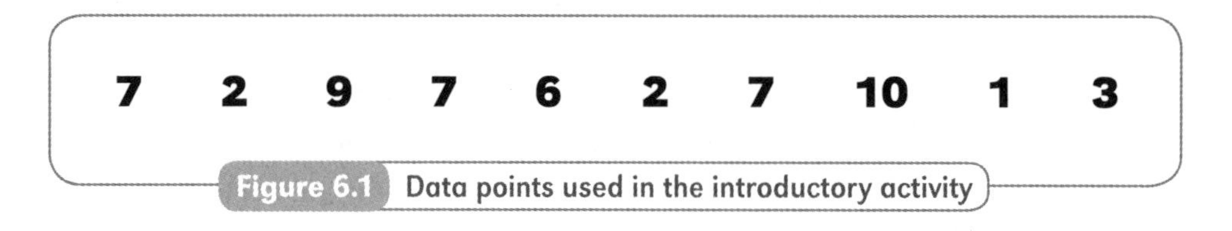

Figure 6.1 Data points used in the introductory activity

The teacher notices the students at Kelvin's table working well together to solve the problem, so she asks Kelvin to share his group's thinking. Kelvin says, "Marla thought that it would be a good idea to first put the data in order. While we were doing that on our whiteboard, Yanni said he knew that 6 would have to be listed as the third number of the five numbers, because the median is always the middle number. So, we found the place in the data where the 6 was the third number and it's 2, 3, 6, 7, 7."

The teacher asks the class if there are any questions for Kelvin. One student, Sarah, said, "I don't really have a question, but when you were saying what you did, I was thinking that I can pretty easily find the median for any five data points if the numbers are all in order. We were just trying to move numbers all around, but putting the numbers in order first would have really helped." Tom then says, "Wait! That means we can just put the six in the middle and then choose two numbers less than six and two numbers greater than six. There are many possible solutions."

The teacher is pleased that this introduction has provided her students with the opportunity to discuss efficient strategies as well as learn more about medians. As she relates what they just did to what they will do when they play the game, she is thinking that she will watch carefully for how her students apply these ideas.

Tips from the Classroom

- We recommend caution in stepping in to assist students if, at first, they make ineffective choices about where to put their numbers. When we field-tested this game, we noticed that, as students played, they developed better strategies for recognizing the relationships among the data and the mean, median, and range.

- You may wish to give some students access to calculators to check their computations of the mean.
- If opposing teams tend to make decisions at a very different rate, you may wish to switch teams or impose a time limit on how long each team has to record the data point.

What to Look For

- Are students using accurate procedures to find the statistics? If not, what misconceptions do they have?
- Are students making wise choices about where to place the numbers? If they are making ineffective decisions despite repeated efforts, discuss what they are thinking as they make their choices in order to better understand their reasoning.
- How are students sharing responsibility for deciding where to assign data points?

Variations

- You may choose to eliminate one of the statistics and replace it with mode or another statistic that you would like your students to explore.
- You could vary the number of data points, keeping in mind that finding the median for an even number of data points requires a different understanding than finding it with an odd number of points.
- You may wish to vary how a team's score is determined. One group recommended earning points based on each team's work. For example, if the target mean were five, the winning team hit the target, and the opposing team had a mean of eight, the winning team would score 8 – 5, or 3, points.
- As you cannot get a range of ten given these numbers, some students suggested adding the rule that if a ten were drawn for the range, it would be replaced with another card.

Exit Card Choices

- If you were able to choose whether the card showing 9 would be the target mean, median, or range, which would you choose? Why?
- What are some ideas that helped you decide where to write a number?

After playing the game, a student provided a response to the first exit card, shown in Figure 6.2. Note that the student's thinking is correct about how many numbers he needs to have on either side of the median, but that he is not necessarily thinking about the probability of getting two numbers greater than nine in relation to getting two numbers less than nine. It would be interesting to have a conversation with this student about the connection that this game has to probability.

I would rather have 9 be the median because I have more control over the median. I would only need two numbers less than 9 and two numbers more than 9.

Figure 6.2 Student response to the first exit card

Extension

Have a group of five (or more) students work together to provide a context for data. In this case, students turn over three cards (one each for the median, range, and mean) and try to find data about themselves that would be closest to each target statistic. For example, students could see how close to the mean they could get by telling how many siblings they each have.

Why This Game or Puzzle?

Lynn Steen, a pioneer in the field of quantitative literacy, wrote that, "The essence of QL is to use mathematical and logical thinking in context," (2004, 47). In *Data Sense* puzzles, students are given numbers to fit correctly into a story about conducting polls. Though the puzzles are challenging, students are often able to increase their tolerance for struggle when they know that correct answers are provided. The puzzlers' goal is to figure out where each number fits.

Two puzzles are provided and both require conceptual understanding of measures of center and spread, deductive reasoning, and the ability to consider what makes sense in each real-world context. Once all the numbers are placed, Puzzle B also requires students to make a recommendation based on the data collected.

***Data Sense* Puzzle A**

Name(s): ______________ Date: ________

Write the numbers in the blanks so that the math makes sense.

Amy was interested in knowing about how many hours a week (to the nearest hour) her classmates play video games on school nights. She surveyed the _____ students in her class (including herself), but she can no longer find her list of the data. She does remember that _____ percent, or _____ students, did not play at all. She knew the range was _____, the median was _____, and the mean was _____. She also found a note she wrote reminding her to tell her mom that she and her best friend were the only ones in her class who play _____ hours per week, or_____ hour less than the mean, and that _____ percent of her class play more than _____ hours per week. Later, she remembered that only_____ classmate reported playing the greatest number of hours. That response was really an outlier, as no one chose the four hourly choices before it.

Fortunately, Amy finally remembered where she wrote her data!

$16\frac{2}{3}$	2
11	3.5
1	30
3	5
50	1
3.5	

Math Focus

- Finding the mean, median, and interquartile range, given a set of data

Materials Needed

- 1 *Data Sense* Puzzle A or B per pair (page A-60 or A-61)
- Optional: 1 *Data Sense* Directions per pair (page A-62)
- Optional: 1 calculator per pair

Directions

Goal: Accurately write the numbers in the blanks in the paragraph.

- Work with a partner.
- Begin by cooperatively reading the paragraph, recognizing the missing numbers as statistical data.
- Interpret the details of the puzzle by using reasoning and sense-making, and calculating the statistics.
- Place the numbers from the box into the blanks. Each number in the box is used exactly once.
- Reread the paragraph to make sure it makes sense.

How It Looks in the Classroom

A teacher introduces his students to the *Data Sense* puzzles by asking them if they like to solve jigsaw puzzles. Many of them say they do, but some say that it is sometimes hard to figure out where to start. The teacher shares that the good news about the puzzle they are going to solve today is that the puzzle already has a start. He says, "The words are already in the puzzle and you have to find where the rest of the pieces—the numbers—go." This comment seems to relieve some of the students in his class who initially seemed concerned about the need to, according to Toby, "Start from scratch."

The teacher displays on the whiteboard a mini-version of a *Data Sense* puzzle, as shown in Figure 6.3. Off to the side of the puzzle, the teacher displays four number cards on the board by putting magnets on the backs of the cards. He asks students to talk at their tables about what number they think belongs in each blank in the puzzle. He is pleased to hear a lively discussion at each table and to learn that the puzzle doesn't seem too easy or too difficult for any one group to solve. He wants his students to feel that working together will help them when solving the more challenging puzzle to come.

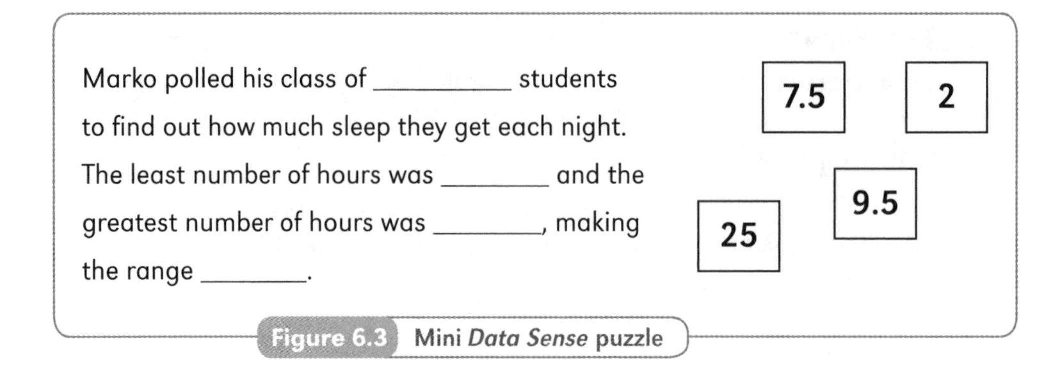

Figure 6.3 Mini *Data Sense* puzzle

When the students come back together, the teacher asks for volunteers to come to the board to put the numbers in the correct places. He allows a group of students to come to the board at the same time, as he knows that some of his students feel more comfortable if they are with a group at the front of the room. Saul's group places the numbers correctly. Then the teacher asks Saul to choose a classmate from another group to provide an explanation for that group's thinking.

Bella is happy to be chosen and she explains that, "Although my group wanted to put 25 in as the greatest number of hours, we realized that if we did, the next greatest number was 9.5 and there aren't 9.5 students in a class. So we made 9.5 the greatest number of hours and 25 the number of students in the class." She goes on to say, "The other numbers just fell into place after that." Then Kathryn comments about her group's thinking. She says, "Our group realized that you can't sleep for 25 hours a night, so that's why we knew right away that the number 25 had to be the number of students in the class."

This teacher is pleased that his students discussed the puzzle using contextual reasoning, rather than just trying to place numbers without thinking about what was realistic. As he sends his students off to try the first puzzle, he explains that this puzzle is more challenging than the one they just did together, but taking their time and talking to each other would help them to reach a solution.

Tips from the Classroom

- We found that some students wanted to start the puzzle alone before they began working with a partner. After working for a bit, though, they tended to want to talk with someone else. For example, we observed two students working side by side, when one turned to the other and said, "I know that there are eighty people buying tickets. So one-eighth of eighty is ten and then I multiplied three by ten and I got thirty. I know $\frac{30}{80} = \frac{3}{8}$ but I don't know what I do with three-eighths." We were particularly pleased when the other student replied with a question, rather than just telling an answer. He asked, "Why do you want to simplify the ratio?"
- Making calculators available may allow some students to give greater attention to the deductive reasoning required.

- You may wish to highlight particular parts of the story, asking students to consider finding these missing numbers first.
- Be sure students use pencils, or put the puzzle inside a sheet protector and use dry-erase markers, so that they can erase their work and begin again when necessary.
- Some students wrote the numbers on small pieces of paper so that they could move the pieces around quickly and easily, when they were considering different options.

What to Look For

- Are there parts of the story that a number of students find difficult, such as thinking about how a change in one piece of data might change the mean? Are they applying what they know about how to find the mean—that is, thinking about how the total of all the data points would change?
- Do students approach the puzzle in an efficient manner?
- Do students check their solutions once they have filled in all of the blanks?
- How successful are students in describing their strategies to others?
- How do students use deductive reasoning and contextual clues to solve the puzzle?
- What did students suggest that Liam should recommend to the coach?

We found that students were engaged in the task of making recommendations. Two student responses are shown in Figure 6.4.

Student A

Liam should tell the coach to make the team practice two hours each day, or 10 hours. It's much more than the mean and the median and its even more than the range. They would win!

Student B

we want our team to be above the mean and to practice more than 50% of the others teams, So we could practice 7 hours each week. we think Liam should suggest eights hours each week. Its' not the greatest number, but the team could practice two hours each days, and have Fridays off.

Figure 6.4 Two students' recommendations

Variations

- Create a puzzle with less data to fill in, making the puzzle easier to solve. You may want to make a puzzle that focuses on one aspect of statistics, such as median or mean.
- As a method of making this puzzle accessible to struggling students, provide a few numbers in the puzzle to get them started.

Exit Card Choices

- Which numbers did you find easier to place? Why?
- Janet collected data about the average number of hours each week that students practiced playing their musical instrument. After she asked three students, the mean for the data was 5 hours. After Janet asked one more student, the mean changed to 4 hours. How many hours per week did that student practice?

Extension

Have students return to the first puzzle but, in this case, do not include the numerical data to write in the blanks. Instead, have students create their own set of numbers that will fit in the puzzle. Or, have students create a possible data set for Amy's *original* data that fits the information in their completed Puzzle A.

Why This Game or Puzzle?

In eighth grade, students investigate patterns of association between two variables. Such explorations often require students to organize data and to apply probabilistic thinking. For example, categorical data might be organized in a table. Students could then consider questions about the data, such as: *Do sixth and eighth graders choose different drinks at lunch?*

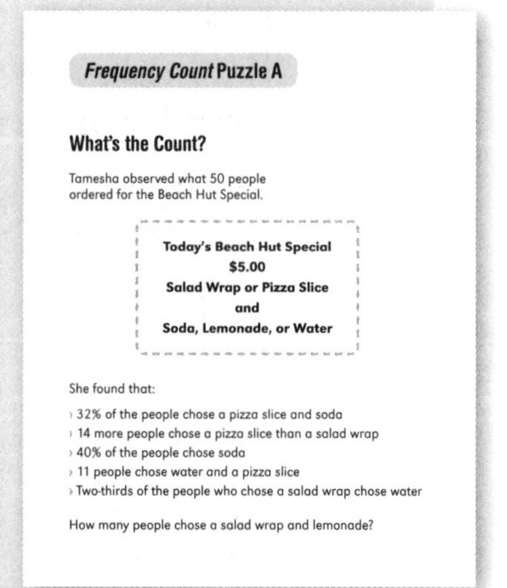

Frequency Count Puzzle A

What's the Count?

Tamesha observed what 50 people ordered for the Beach Hut Special.

Today's Beach Hut Special
$5.00
Salad Wrap or Pizza Slice
and
Soda, Lemonade, or Water

She found that:

- 32% of the people chose a pizza slice and soda
- 14 more people chose a pizza slice than a salad wrap
- 40% of the people chose soda
- 11 people chose water and a pizza slice
- Two-thirds of the people who chose a salad wrap chose water

How many people chose a salad wrap and lemonade?

Eighth graders could make a frequency table to solve this puzzle, while many seventh graders and some sixth graders could solve it through proportional reasoning. In high school, students will build on this thinking as they learn to define weak and strong associations through correlation coefficients (Common Core Standards Writing Team 2011b).

The *Frequency Count* puzzles do not tell students how to use the information to respond to the questions posed. Rather, students need to think like mathematicians—that is, to consider the data, decide how to organize it, and draw conclusions.

Math Focus

- Investigating patterns of association given data about two or more populations

Materials Needed

- 1 *Frequency Count* Puzzle A or B per group (page A-63 or A-64)
- Optional: 1 *Frequency Count* Directions per group (page A-65)
- Optional: 1 calculator per group

Directions

Goal: Answer the question(s), as a result of interpreting the information provided in the puzzle.

- Work alone or with a partner.
- Read the clues.
- Make notes so that you can recall your thinking.
- You do not need to use the clues in order.
- Check your solution with each clue.

How It Looks in the Classroom

As one teacher prepares to introduce the *Frequency Count* puzzle to her students, she realizes that some might need a refresher on finding percentages and fractional amounts of a whole. So she begins her class by polling the students as to how many prefer using mechanical pencils. She then asks if they like using lined or graph paper to do their work. As the results are recorded, the teacher asks the students to find the percentage of students in the class who prefer mechanical pencils, then the percentage who prefer lined paper. Lastly, she asks them to find the percentage who prefer both and the percentage who prefer neither.

Sal is pleased to note that he doesn't need to calculate the number of students who preferred neither. He tells the class, "I just added up those who liked using mechanical pencils as forty-five percent and those who liked using lined paper as fifty percent, got ninety-five percent, and decided that those who don't like either must be five percent." Calvin says, "Sorry, Sal, but I don't think you are right. You forgot about the fact that some of the students fit into both categories." As further discussion ensues, with different strategies proposed for determining the percentage of students who preferred neither, the teacher is pleased that this misconception came about naturally as it will be important for students to understand as they work through the puzzle.

The teacher then turns the students' attention to how the puzzle is different from, but related to, the activity the students have just completed. "In this puzzle, you will be given the percentages and some fractional amounts of the total, and you will need to determine how many belong in each category." A few students are overheard saying that the puzzle seems like it will be a lot harder, but that maybe what they just did together will help.

Tips from the Classroom

- You may wish to provide some students with a template, such as a chart, for recording their data. Eighth graders may benefit from being provided a blank frequency table in which they can record their solution.
- Encourage students to use mental math to find the percentages, and only after doing so, use a calculator to check their work.
- Some students wanted to try to answer the question(s) without finding and organizing all of the data, but later found they had difficulty solving the puzzle. Consider requiring such students to show the information in a chart as a strategy for answering the question(s).
- Some students may need advice about where they could most efficiently start in the list of clues.

What to Look For

- Are students persevering through the puzzle? Do they need a reminder that puzzles take time to solve and that it's okay if they can't solve it right away?
- Do students compute accurately and with conceptual understanding?
- Are some students able to do the mathematics but having trouble interpreting the wording in the clues?

Variations

- Put each of the clues on a card, hand one card to each student in the group, and have students reveal one card at a time to solve the puzzle.
- Consider the different levels of learners in your classroom and include some additional, or different, questions at the end of the puzzle.
- Create a new, simpler puzzle with only two choices for each category.

Exit Card Choices

- Which clue was the most helpful to you in solving the puzzle? Explain your reasoning.
- Your partner said, "Thirty-two percent of 50 is 16 because 32% of 100 is 32 so I just need to take half of 32." Do you agree with this reasoning? Why or why not?

Extension

Have students write a tip sheet for other puzzlers who have not yet solved the puzzle, providing ideas for how to organize the information or how to use deductive reasoning.

I'll Take My Chances

Why This Game or Puzzle?

This game provides a motivating environment in which students can explore the likelihood of a particular outcome when simulating an event. Players are given a deck of playing cards, two dice, and two coins. They are also given outcomes, such as *A multiple of 2 when picking one card from the deck* or *A number less than 4 when rolling a die or heads when tossing a penny.* Teams use the cards, dice, and/or coins to attempt to meet the outcome, repeating the experiment as many times as necessary and aiming for fewer trials than their opposing team. After six outcomes are explored, the team with the fewer number of trials wins.

I'll Take My Chances Card Set

A Both coins land on heads when tossing two pennies	B A multiple of 2 when picking one card from the deck
C A sum of 6 when rolling two dice	D A 3 or a 4 on either die when rolling two dice
E A black card that is less than 5 when picking one card from the deck	F An odd number on each die when rolling two dice
G At least one penny lands on heads when tossing two pennies	H A red card and a black card when picking two cards from the deck

It is important that students have opportunities to replace their faulty ideas about chance with ideas based on objective observations (Shay 2008). There are sixteen cards listing outcomes, which will support students playing the game several times. The quick pace of this game holds students' attention and lets them gain an intuitive sense of which conditions are more challenging to meet than others.

Math Focus

- Investigating chance of simple and compound events

Materials Needed

- 1 standard deck of cards, 2 dice, and 2 pennies per group
- 1 *I'll Take My Chances* Card Set per group (pages A-66 to A-67)

- 1 *I'll Take My Chances* Recording Sheet per team (page A-68, top or bottom half)
- Optional: 1 *I'll Take My Chances* Directions per group (page A-69)

Directions

Goal: Get all six outcomes with the fewer number of trials.

- Place all the *I'll Take My Chances* cards facedown between the two teams.
- Choose six cards to use for the game and mark the letter of each card chosen in the same order on each team's recording sheet.
- Take turns.
- On its first turn, Team 1 reads the situation written on the card and both teams agree on the expectations. Team 1 then carries out the simulation.
- If the team is able to match the event in one try, it checks it off on the recording sheet. On its next turn, Team 1 will try to match the event on the next card.
- If the team cannot match the event, its turn ends and the team makes a tally mark on its recording sheet to indicate that it tried the event. Team 1 must try again to match the event on its next turn.
- The teams then alternate turns, trying to match the outcomes described in the order they appear on the recording sheet.
- The first team to match all outcomes wins the game.

How It Looks in the Classroom

As the students enter one seventh-grade classroom, they notice there is a deck of cards, two dice, and two pennies on each group's table. The students are interested in knowing what this day's probability lesson will entail. As the teacher draws their attention to the board, they notice some probability outcomes written on cards and the teacher overhears several students comment that they hope they will be doing experiments today.

As the teacher knows the students are eager to use the manipulatives, she reminds them to wait for directions to do so. She then asks Robert to choose a card from the three that are displayed and he chooses the card that reads *Toss a coin twice and get heads both times.* The teacher then tells the students to toss one of their pennies two times and, if they get heads each time they toss it, to raise their hands. The teacher isn't surprised when she sees several groups tossing the coin more than two times, but she knows she has to make sure the expectations are clear. She chooses to stop the class and ask Balik to restate what is written on the card. He reminds the class that it said to toss the penny two times. As heads nod, one student says, "What if we don't get heads twice?" The teacher comments, "Let's first see how many of you get heads two times. Then you can toss again, keeping track of how many times it takes you to toss one penny twice and get heads two times in a row."

After each group has successfully reached the desired outcome, the teacher asks the students how many times it took them. One group reports that it took them four tries, while another says it took them only two. Nieve's group reports that it took them nine tries, and Nieve says, "It seems like it shouldn't take that long. Maybe you gave us bad pennies." The teacher is pleased to hear her thinking about how many tries it should take them to achieve a desired outcome and is looking forward to hearing more as her students start playing the game.

Tips from the Classroom

- If you think that it would work better for the students in your class, provide each team with their own set of cards, dice, and pennies.
- During our field-testing, we recognized that some students benefited from having their own copy of the cards that they could place or paste on their recording sheets.

What to Look For

- What conversations do you hear about the likelihood of an event?
- Are students clear about the difference between *and* and *or*?
- Are teams paying attention to the accuracy of the other team's outcomes? Are teams waiting until the opposing team has finished before taking their turn?

Variations

- Students may want to order their cards differently on each of their recording sheets. In our field-testing, we found that doing so made for some interesting conversations about why they would want a particular card first or last.
- You may want to give students the opportunity to make their own sets of cards to add to the deck.
- You could have teams conduct the experiment until the expected outcome occurs. In this case, teams would keep track of the number of trials required, and the team with fewer total trials would win that round.

Exit Card Choices

- Why do you think some outcomes are harder to get than others?
- Would you rather have a card that has you getting an odd sum when rolling two dice or a card that has you selecting a black card when picking one card from the deck? Why?

Extension

Have the students place the cards in order based on the number of trials they think are likely to be needed, from least to greatest, to achieve each outcome. Have students write about why they chose the order that they did.

Is It a Match?

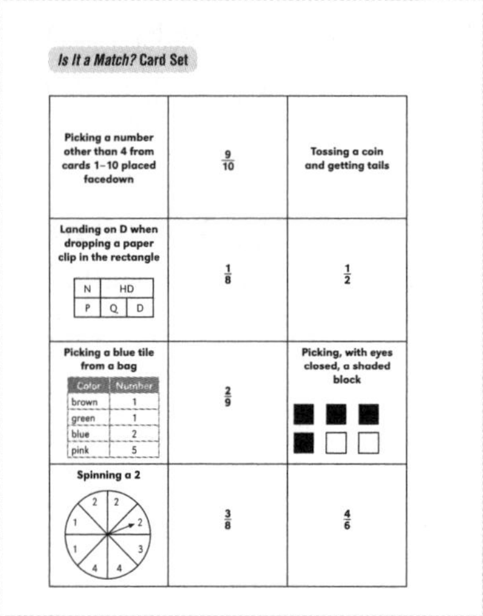

Is It a Match? Card Set

Picking a number other than 4 from cards 1–10 placed facedown	$\frac{9}{10}$	Tossing a coin and getting tails
Landing on D when dropping a paper clip in the rectangle (N, HD / P, Q, D)	$\frac{1}{8}$	$\frac{1}{2}$
Picking a blue tile from a bag (Color / Number: brown 1, green 1, blue 2, pink 5)	$\frac{2}{9}$	Picking, with eyes closed, a shaded block
Spinning a 2	$\frac{3}{8}$	$\frac{4}{6}$

Why This Game or Puzzle?

According to Koellner and her colleagues (2015), it can be difficult for students to understand probability. They need ample opportunities to talk about their thinking and consider a variety of events and their outcomes. *Is It a Match?* focuses on theoretical probability, rather than experimental probability. It is a concentration game in which students try to match the description of an event with the probability of its occurrence. The game is designed to reinforce the idea that the likelihood of an event is reported as a fraction, with the numerator telling the number of favorable outcomes and the denominator indicating the total number of possible outcomes. As students determine the fraction representation, they may recognize that the greater the number, the more likely the event.

Math Focus

- Understanding that the probability of an event is represented by a fraction
- Finding probabilities of simple and compound events

Materials Needed

- 1 *Is It a Match?* Card Set per group (pages A-70 to A-72)
- Optional: 1 *Is It a Match?* Directions per group (page A-73)

Directions

Goal: Get the most card sets when all cards have been matched.

- Shuffle all the cards and place them facedown in a 6-by-6 array between the two teams.
- On each turn, the team:
  1. Turns over two cards.
  2. The two cards match if one card represents an event and the other

card gives the probability of its occurrence. If the cards match, the team keeps the cards and turns over another two cards.

3. If the cards do not match, the team must return the cards facedown in the same place they were in the array.

- The game is over when all the cards have been matched.
- The winner is the team who has the most matched card sets.

How It Looks in the Classroom

To introduce this game, one seventh-grade teacher places the *Is It a Match?* cards that describe events in a bag and asks Fatima to choose one and read out loud what is written on the card. Fatima reads, "Tossing a coin and getting tails."

The teacher has placed the *Is It a Match?* cards that show probabilities in another bag and asks Fatima to choose a card from this second bag and again read it aloud. Fatima tells the class that the card shows $\frac{1}{4}$. The teacher asks the students if they think that the fraction $\frac{1}{4}$ represents the probability of getting tails when a coin is tossed. Barry responds, "I think that it's possible to be $\frac{1}{4}$ because there is only one way to get a tail." Jessie follows by commenting, "Yes, that's true, but there aren't four sides to a coin, so there aren't four different possibilities—so, $\frac{1}{4}$ doesn't work."

The students take turns drawing cards from the second bag, until they find $\frac{1}{2}$. A number of students said in unison, "That's it!" Discussing what the $\frac{1}{2}$ represents, Meek suggests, "I think the $\frac{1}{2}$ means that half of the time that you toss a coin you should get tails." The teacher is pleased to hear Meek share his answer in the form of a fraction, rather than distinct whole numbers, so that his classmates can connect theoretical probability to a fraction between zero and one. She also wants students to remember the difference between experimental and theoretical probability. She asks, "So, if we toss the coin two times, will we get one tail?" Violet says, "Not always, but if we tossed it many times, we would get tails about half the time. It's like a theory."

As the teacher explains the rules of *Is It a Match?*, she encourages students to keep in mind that the game requires them to think about the probability of an event, as well as to keep track of where cards that have been previously turned over are found in the array.

Tips from the Classroom

- You may wish to copy the cards onto heavier, colored paper so that students are unable to see through them when they are choosing which ones to turn over.
- Encourage students to give each other enough time to see what is on the card as it is turned over, prior to turning it facedown again. It may be necessary to have students read the card aloud as a method for slowing them down. Also, the auditory memory can support the visual memory.
- You may need to remind some students to think about equivalent fractions as they are considering the given probabilities.

- Some students may wish to simulate the event in order to better understand the probability, so you might want to have coins, cards, and dice available when they are playing the game.

What to Look For

- Are there students who cannot find a match when the fraction shown on the card is equivalent to, but different from, the one that they found? Encourage students to look for other forms of their answer when considering whether or not a card is a match.
- Do students more easily grasp probability with one type of situation than another? As a follow-up, we encouraged those who seemed to better grasp probability with dice to engage more in experimental activities that required them to use cards, coins, or spinners, so that they became better accustomed to those events and outcomes.
- Are some students able to do the math, but unable to remember where matching cards were placed?

Variations

- As a way of scaffolding the game, allow some students to start the game with some of the cards face up.
- In our field-testing, some students decided that they wanted to change the rules so that teams who found a match did not immediately get another turn. We encouraged students to make rules that worked for them, as long as all players agreed and the mathematics was not impacted.
- You may wish to play with more or fewer cards.

Exit Card Choices

- Using two dice, create a situation in which the probability of the event occurring is $\frac{1}{4}$.
- A student created a spinner to play a game, and the likelihood of spinning a 2 on the spinner was $\frac{1}{3}$. Draw two spinners that the student could have created.

Extension

Have students choose some of the matching cards and create a simulation—for instance, picking cards or tossing coins. Support them in understanding the difference between theoretical and experimental probability, perhaps by using an app or computer program that would generate numerous trials.

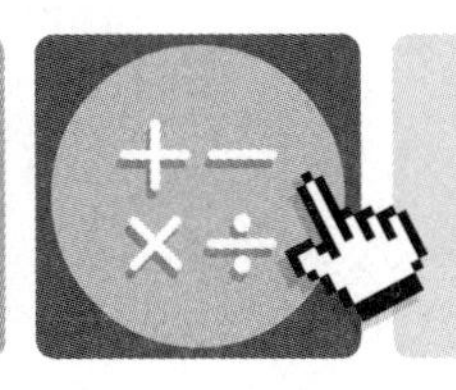

Online Games and Apps

Students certainly benefit from using online statistics and probability simulations to learn more about these topics. Many such activities give students the opportunity to collect large amounts of data through random number generators, simulated spinners, dice, or decks of cards. Often web-based applets also provide information about the experimental probability data collected and the related theoretical probability.

Of particular interest to students are those online games and apps that allow them to manipulate data to determine measures of center and variability, draw conclusions about populations, and create their own probability models. A few examples include:

- The free game, Landmark Shark (http://media.emgames.com/emgames/demosite/playdemo.html?activity=M5A006&activitytype=dcr&level=3), randomly provides five cards to the player, who subsequently must choose whether the greatest statistic— the "landmark" in the game—is the mode, median, or range. If the correct landmark is chosen, the players are awarded that number of points. If players choose to do so, they can exchange up to three of the number cards in an attempt to find a greater landmark. For additional points, students compute and input the mean of the data set.

- Run That Town (http://runthattown.abs.gov.au), a free app, begins by allowing players to be the mayor of a real town (of their choice) in Australia. A simulation game, players make decisions about town projects using real Census data, and must consider how to use that data in conjunction with public opinion. Proposals are accepted or rejected for such projects as building cafés, constructing sports arenas, and creating town parks. The mayor becomes more or less popular based on the satisfaction level of the town's residents.

- Stick or Switch (free from http://illuminations.nctm.org) engages players in a game-show scenario in which they choose from one of three doors in an attempt to win a prize. One of the non-winning doors is opened and players are given the choice to stick or switch. Engaging players in thinking about the best chances of winning, this game is an online version of the "Monty Hall problem," a classic puzzle named after the first host of the game show, *Let's Make a Deal.* Players may choose to play the game using guess-and-check, experimentation, simulations, or theoretical models.

Patterns, Graphs, and Functions

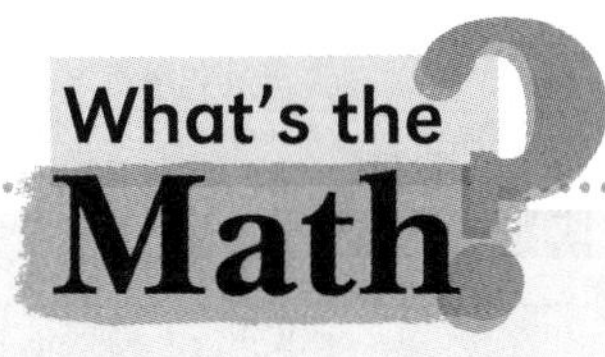

Ideas related to patterns, graphs, and equations permeate the middle school curriculum. Students in grades six and seven continue to develop their conceptual understanding of patterns, graphs, and algebraic thinking as they build a foundation for the study of functions. Students in eighth grade begin an initial study of functions, which is extended and formalized in high school. In this chapter, we have included some games and puzzles that sixth and seventh graders as well as eighth graders can explore.

When students recognize patterns between numerical data in an input-output table, they are developing an intuitive understanding that the output value (y) is a function of the input value (x). Most important, they are exploring, predicting, and generalizing about quantities that vary (Star and Rittle-Johnson 2009). Students learn to express the pattern in an equation and/or as ordered pairs in a graph, later recognizing that functions assign exactly one output to each input.

In eighth grade, students learn that a function may be represented by:

- a description
- a table
- a graph
- an equation

We want all students to be fluid with these multiple representations (Van de Walle, Karp, and Bay-Williams 2013), be able to make connections among them, and gain a sense of when each is most useful in solving problems.

Eighth graders should also recognize functions as linear or non-linear. They learn that an equation in the form $y = mx + b$ is a linear function, can find the rate of change given two points, and learn to interpret the axes' intercepts in the context of a real-life situation. These students also can recognize when functions are increasing or decreasing, at a constant or non-constant rate, and work flexibly among descriptions, graphs, tables, and equations.

Patternary

Why This Game or Puzzle?

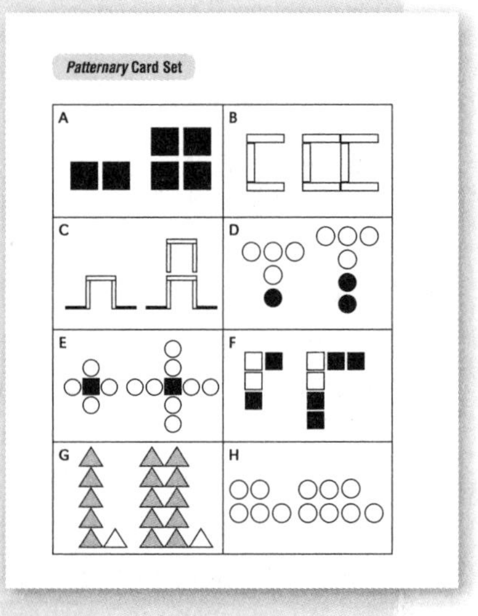

Jo Boaler (2013) makes a distinction between procedural algebra and structural algebra. She suggests that structural algebra is supported when we ask students open-ended questions about growing patterns. The open nature of this game gives students the opportunity to expand their sense of structural algebra. It also represents how researchers work—that is, they observe patterns and then try to discern potential relationships among the data. This type of thinking is quite different from being told to identify a relationship between defined variables *A* and *B*.

In this game, the drawing team is given a *Patternary* card of a growing pattern. The card shows the first two images, without independent and dependent variables identified. The players consider the images, decide on the growing pattern they wish to use, and then make the next four images in the pattern. They record these decisions on their recording sheets. They show the card and then these drawings, one at a time. The guessing team tries to guess the pattern, either by describing it and/or by writing an equation to represent it. Both teams receive points based on whether they identify the correct pattern and the number of tries needed to do so. The game is designed in this cooperative manner so that both teams are interested in communicating as clearly as possible.

Math Focus

- Identifying, extending, and describing a growing pattern
- Identifying a relationship between two quantities in a diagram
- Using variables to represent numbers when modeling a real-world situation

Materials Needed

- 1 *Patternary* Card Set per group (pages A-74 to A-75)
- 1 *Patternary* Recording Sheet per team (page A-76)
- Optional: 1 *Patternary* Directions per group (page A-77)

Directions

Goal: Play cooperatively to earn a total of 40 points by identifying and continuing growing patterns.

- The *Patternary* Card Set is shuffled and placed facedown between the two teams.
- For each round of the game, each team:
  1. Chooses a card from the card set, considers the images on the card, and thinks about how the pattern might be growing;
  2. Identifies a growing pattern and continues it by drawing the next four images on the recording sheet;
  3. Writes an equation or a description to represent the growing pattern.
- Team 1 shows Team 2 players the *Patternary* card it chose. After seeing the card, Team 2 gets one try to provide a description or an equation (or an equivalent description or equation) that matches the growing pattern Team 1 identified. If Team 2 cannot do so, players ask to see the first image Team 1 drew to extend the pattern.
- Play continues with Team 1 showing one drawing at a time and Team 2 being given one try to identify the pattern each time a new drawing is shown.
- If Team 2 is able to provide a description or identify the equation by the time the fourth student-drawn image is shown, both teams earn 15 points, minus 1 point for each drawing shown. If Team 2 is unable to identify the equation after seeing the original card and the four drawings, it is asked to make a drawing that fits the pattern. If Team 1 agrees that the drawing is correct, both teams earn 5 points.
- If one team has recorded a description and the other an equation, teams talk until they agree there is a match.
- Teams reverse roles and complete the round.
- After two rounds, both teams are considered to be winners if the group has earned 40 points.

How It Looks in the Classroom

This seventh-grade teacher made a card similar to the *Patternary* cards (Figure 7.1), showing the first two images in a growing pattern. He projects the pattern card to the class, asks what pattern the students see, and records their ideas on the class conjecture board. He is pleased with students' enthusiasm for the task and the number of observations they made (Figure 7.2).

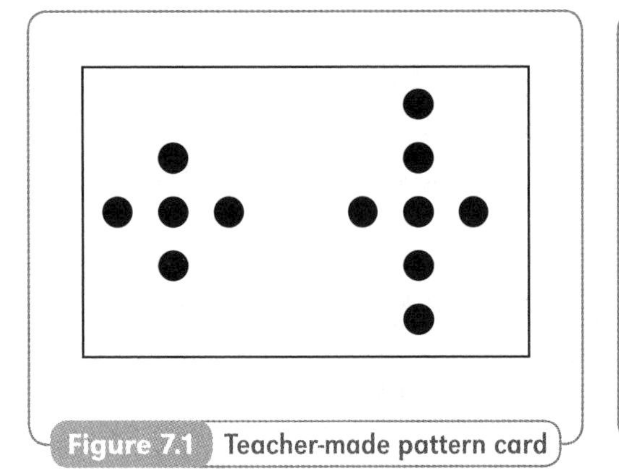

Figure 7.1 Teacher-made pattern card

- The up-and-down part is growing.
- All of the shapes are circles.
- The total number of shapes in each design is odd.
- There is one more on the top and the bottom of the second one.
- Both have three circles across the middle.
- If you look at the circle in the middle, there is always the same number above and below it.

Figure 7.2 Recording on the conjecture board

The teacher asks, "Do you know what the next one will look like?" Zarah is certain the next image will show three circles across and four circles on the top and the bottom, saying, "The middle always stays the same and the number on the top and the bottom doubles." Bobbi is just as certain that there will be three circles across and three on the top and the bottom, as she thinks there is one added each time. David believes the row of three will start growing next. Eric says, "There is not enough information yet," which prompts the teacher to project the next drawing, beside the original card, as shown in Figure 7.3. He tells the students to talk with their partners about the growing pattern they see and to try to describe it in words or write an equation to represent it.

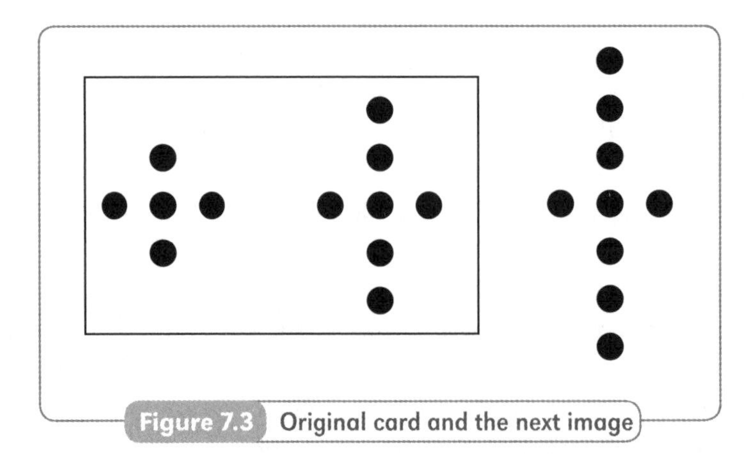

Figure 7.3 Original card and the next image

After some time to reflect, Paul reports, "We think that it's $t = 3 + 2n$." The teacher asks the partners to state why they think that this is the right equation and to explain the meaning of their rule. Malik speaks for the group and says, "We are looking for the total number based on the number of the drawing. We see three in the middle, so that's plus three. Then there are the same number of circles on the top and bottom as the number of the drawing, so we multiplied by two."

Randy says, "I can understand that, but we saw it differently. We wrote *The total number is the two that stay the same each time, then twice the number of the drawing, plus the one in the middle.*" Kerry says, "There are two ways to see it, but the totals would be the same. I think we call that equivalent."

The teacher asks other students to make comments or ask questions about these ideas. Sophie says, "I don't know what you mean by the two that stay the same each time." Evan asks if he can point to them on the projected image and indicates the circles on the left and right of the middle one. Other students seem pleased that Sophie has asked that question; they are nodding their heads and saying they now understand the "two" Randy was talking about and why they needed to add the one at the end.

The teacher explains the rules to *Patternary*. As the students gather in smaller groups to play the game, several are overheard discussing how they will decide what their patterns will be. The teacher is pleased that the game has already motivated students to talk about their strategies.

Tips from the Classroom

- It may be easier for some students if Team 1 identifies what it considers to be the independent and dependent variables when it shares the card it chose.
- When we were field-testing this game with students who were asked to write an equation, a number of them accurately drew the images but were uncertain how to write an equation to represent the pattern. You may wish to ask such students to first consider what the tenth or twentieth image would look like by drawing a sketch, as this often leads students to a generalization and then an equation.
- Some students benefited from color coding the patterns, to highlight the parts they saw as staying the same and the parts they saw as changing from one image to another.

What to Look For

- Is a team's ability to continue patterns related to its ability to identify them?
- Are Team 2 students writing descriptions or equations, or showing the next image in the pattern? Are they translating correctly between descriptions and equations?
- Do teams recognize when different descriptions and equations are equivalent?
- Are teams seeing more than one potential pattern? If so, how do they choose which pattern to consider? Are teams sharing the responsibility for choosing and identifying patterns?

Variations

- The growing patterns on the cards could be made more or less complex by involving a different number of shapes or types of shading. You could also cut them so that only the first image is shown, allowing for a wider range of patterns to be generated.

- Students could create their own growing patterns; some may want to use physical models such as pattern blocks or linking cubes to do so.
- Students could be required to use the same card twice, using different variables the second time. For example, for the card shown in the "How It Looks in the Classroom" section, you could use the figure number to predict the number of circles in the vertical column and get the equation $t = 2n + 1$.
- Once students become familiar with the game, you may want to adapt the scoring and allow teams the option of playing competitively.

Exit Card Choices

- If your team chose this card, what four drawings would you make? Describe your growing pattern and/or write an equation.

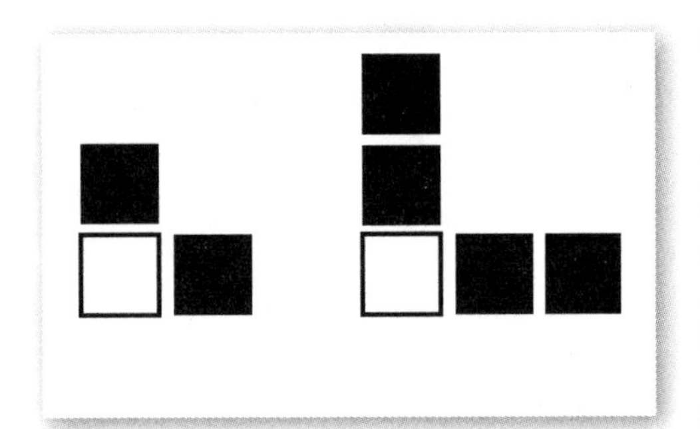

- Create four drawings to represent the function $t = 3f + 1$.

A student response to the first exit card is shown in Figure 7.4. The sixth grader extended the pattern easily and was able to write a correct equation to represent the pattern. However, her explanation needs further details. Would you simply ask for more information, or would you ask a specific question, such as *What does the + 1 represent?*

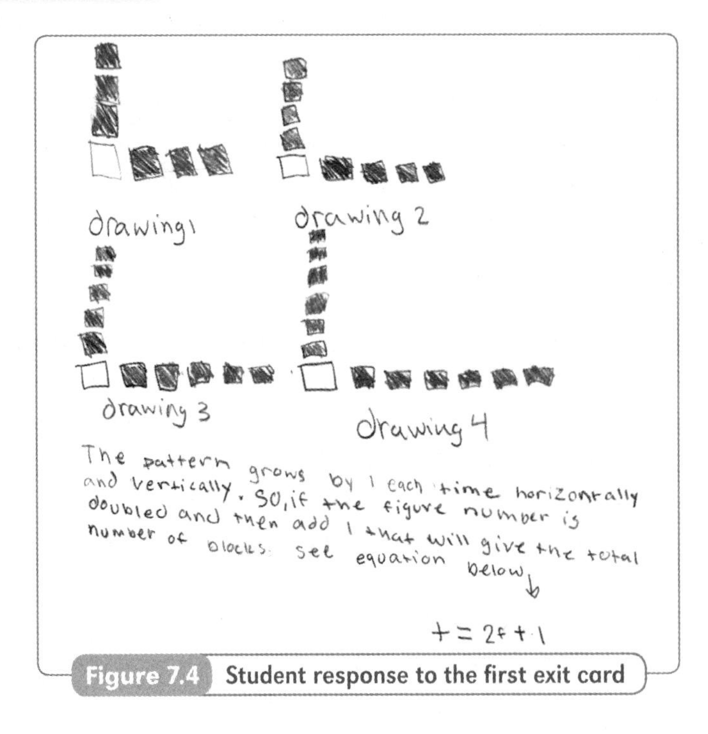

Figure 7.4 Student response to the first exit card

Extension

For a week, start your math class by projecting one image and having students work in pairs to make the next two images. Ask students to share their images and discuss the differences among them. Then have students try to predict the next image in those drawings that are different from their own.

Why This Game or Puzzle?

Algebra gives us a language for investigating and describing patterns (Bressoud 2012). This classic game provides students with opportunities to investigate input-output tables. As they do so, they develop an informal understanding of what it means to have two variables that change in relationship to one another.

In this version of the game, Team 1 is given the rule by choosing one of the equation cards. Team 2 writes an input value in a table and Team 1 determines the associated output value and records it in the table. These interchanges continue until Team 2 is able to identify the rule. The equation rules vary in difficulty and you can choose which cards to offer your students; for example, you can give them the first two rows of cards, the middle two rows of cards, or the last two rows of cards.

What's My Rule? Card Set

$x + 4 = y$	$x + 1 = y$	$2x - 1 = y$
$3x - 1 = y$	$2x + 5 = y$	$2x + \frac{1}{2} = y$
$x^2 = y$	$\frac{1}{2}x - 1 = y$	$2x^2 = y$
$2(x - 1) = y$	$\frac{(x+4)}{2} = y$	$x^2 + 2 = y$

Math Focus

- Evaluating equations in which variables stand for numbers
- Exploring rules in which each input results in one unique output

Materials Needed

- 1 *What's My Rule?* Card Set per group (page A-78)
- 1 *What's My Rule?* Recording Sheet per team (page A-79)
- Optional: 1 *What's My Rule?* Directions per group (page A-80)

Directions

Goal: Identify rules by using fewer input values than the opposing team uses to identify its rules.

- The *What's My Rule?* cards are shuffled and placed facedown as a deck between the two teams.
- Team 1 chooses a card from the deck and keeps it private, as the card identifies the rule for the turn.
- Team 2 gives Team 1 a number to input (x) into the rule and then Team 1 tells Team 2 the output number (y).
- Team 2 tries to guess the rule, in the form of an equation.
- Team 2's score is the number of input values it names before identifying the rule.
- It is then Team 2's turn to choose a card from the deck.
- At the end of three turns for each team, the team with fewer points is the winner.

How It Looks in the Classroom

One sixth-grade teacher knows that her students are familiar with this classic game, but also realizes that they have never played it with the use of the terms "input" and "output" to describe the numbers involved or with the expectation to identify the rule with an algebraic equation. The teacher introduces the game in the following manner to the whole class.

The teacher draws a card from the *What's My Rule?* card set and asks for a volunteer to choose the first input number. Anya chooses seven and one student remarks, "That's not what I would pick." After the teacher records the output for an input of seven, another student comments, "It doesn't matter what number we pick first. Let's just pick another number to help us figure out what the pattern is."

As other students offer number inputs, the students draw closer to what they consider to be the rule, but it is Zac's input that helps them to reach their final conclusion. He said, "I want to choose zero as an input. That way we will know more about where the pattern starts." After a few more guesses Nari is ready to state the rule, which she says is to multiply the input by four and then add two. The teacher comments that they have worked hard to find the rule using words, but the game requires them to identify the rule as an equation. The teacher tells all the students to stop and think for a moment and then they agree that the rule is $4x + 2 = y$.

The teacher then explains the scoring rules of the game and the students are excited about this additional component, which encourages them to identify the rule with as few guesses as possible.

Tips from the Classroom

- Some students may need assistance with translating the written description into an algebraic equation. You may want to ask them to write the description several times, with the inputs included, to see the pattern. Then they can replace the inputs and outputs with variables.
- Correct outputs are essential for the success of the game. Encourage each team to check its output carefully, before responding. You may want to allow students to use a calculator as needed.
- Remind students that teams may identify an equivalent equation, which should be accepted.

What to Look For

- How are students communicating their algebraic thinking?
- What strategies are players using to identify the rule?
- Are students choosing input values strategically or randomly? For example, are they choosing values in order, or suggesting zero as an input?

Variations

- You might require a team to give an input and guess the correct output, before guessing the rule.
- Some students may not be ready to identify an equation for the rule. You may wish to allow them to provide a written expression.
- You could have the guessing team give the outputs and be told the inputs.
- As an alternative to having students choose cards from a deck, they could create their own rules.

Exit Card Choices

- When you gave the input 4, the output was 17. When you gave the input 6, the output was 26. What do you think the output will be for 10? Explain.
- You are making a table as you play the game and it looks like the one shown here. What do you think the rule is? Why do you think so?

Input (x)	Output (y)
6	41
2	13
4	27

Extension

Have students write a journal response to the following prompt:

What strategies do you use to identify the rule in the *What's My Rule?* game?

Graph Story

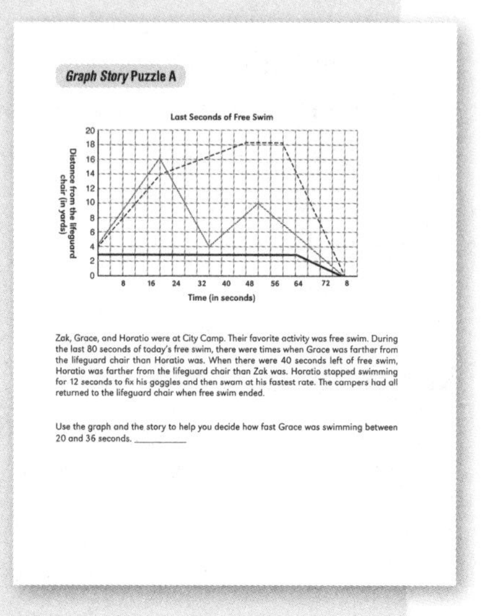

Graph Story Puzzle A

Zak, Grace, and Horatio were at City Camp. Their favorite activity was free swim. During the last 80 seconds of today's free swim, there were times when Grace was farther from the lifeguard chair than Horatio was. When there were 40 seconds left of free swim, Horatio was farther from the lifeguard chair than Zak was. Horatio stopped swimming for 12 seconds to fix his goggles and then swam at his fastest rate. The campers had all returned to the lifeguard chair when free swim ended.

Use the graph and the story to help you decide how fast Grace was swimming between 20 and 36 seconds. ________

Why This Game or Puzzle?

In this puzzle, students are provided with a graph and a story, which represent the time-distance scenarios of three (Puzzle A) or four (Puzzle B) campers. Puzzlers use the given information to identify the part of the graph that represents each camper and then answer a question about the speed of one of the campers. To do so, puzzlers must analyze the story for information about what happened and then find matching features in the graph, or begin with features in the graph and match them to the story. Students might ask themselves questions, such as *Did this camper move away from or toward the end of the dock? What does a line segment that is horizontal to the x-axis tell me? How do I find how much time has elapsed?*

As students work together to solve a *Graph Story* puzzle, they will develop common language to describe the graph. Kara Jackson and her coauthors (2012) identify the development of common language to describe key features as one of four ways to increase students' engagement and success with complex tasks.

Math Focus

- Relating information shown in a graph to a real-world situation
- Solving real-world problems by finding distances between points in a coordinate plane
- Identifying the rate of change in a situation from a graph

Materials Needed

- 1 *Graph Story* Puzzle A or B per group (page A-81 or A-82)
- Optional: 1 *Graph Story* Directions per group (page A-83)

Directions

Goal: Match each part of the graph with one camper's story and answer the related question.

- Read the story as it relates to the graph or look at the graph and relate it to the story.
- Determine which part of the graph matches which camper's story.
- Answer the question.
- Be sure to check your work.

How It Looks in the Classroom

A seventh-grade teacher introduces *Graph Story* to her students by engaging them in the interpretation of a similar graph. She displays the graph and story shown in Figure 7.5. She asks the students to talk with their table groups about whether or not the words and the graph match.

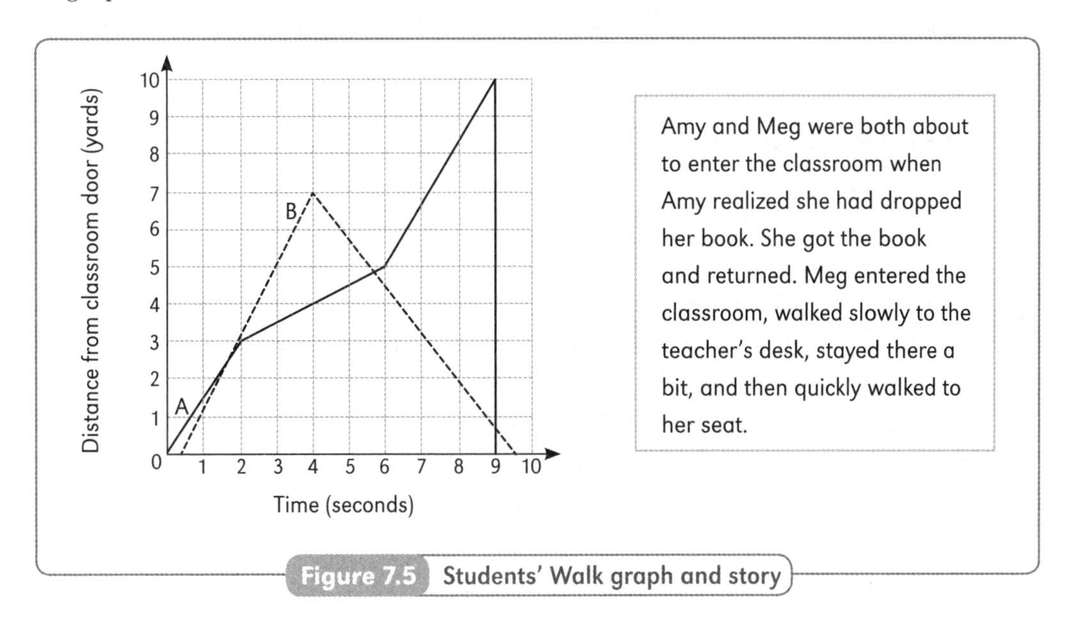

Figure 7.5 Students' Walk graph and story

One group asks if it can use a calculator to determine the exact rate of change for each part of the graph. The teacher waits before responding so that the students have the opportunity to answer their own question. Artem responds, "Wait, we might not need exact rates of change. We just need to be sure to pay attention to how close Meg and Amy are to the classroom door."

At the next group, the teacher hears Cheryl comment, "I think that the line labeled B could be Amy because she gets farther away from the classroom door and then closer. That probably shows her going and coming back with her book."

When the teacher brings the students' attention back to the whole group, most share that they believe that Meg's story is represented by A and Amy's by B. The students refer to distances from the door as they do so. Then Andre says, "I agree with the Amy part, but the description of what Meg did doesn't match the graph." The teacher tells students to look at the description and graph again and think about whether they agree or disagree with Andre's statement.

After some time to review the data, some students disagree with Andre, commenting that both the description for Meg and graph A have three parts to them. Then Portia says, "But her distance from the teacher is changing as she talks to him, and that would be rude." Jerek adds, "I agree—the line should probably show the distance from the door staying constant, but Meg could be moving while she talked, so she would be in her seat when class started. Wait, no, it says she stopped." Lisa then asks if she can show something on the diagram and then points to the first and last segments of A. She talks about the steepness of each line and the language in the story, arguing that the slower walk should be less steep than the quicker one. After a few more comments, the teacher describes the *Graph Story* puzzle and tells students to keep this discussion in mind as they solve it.

The teacher notes that the students were actively involved in the discussion of the relationship between the story and the graph, used relevant vocabulary, and demonstrated an emerging knowledge of connecting stories to graphs. Based on the conversation, she will give some students Puzzle A and some students Puzzle B to solve.

Tips from the Classroom

- For some students, seeing all the campers' stories represented on one graph is challenging. You may suggest that students trace each camper's path with a different-color highlighter or pencil.
- It may be necessary to remind some students about the labels for the *x*- and *y*-axes' scales. Our work with some puzzlers revealed that some just counted the spaces without thinking about a graph's scale.
- Some students may not be familiar with free swims, goggles, or kayaks. Lack of familiarity with context can limit students' problem-solving success. Begin with a brief conversation about each term, as appropriate for your students.

What to Look For

- Do some students appear to have a preference for the information that is given verbally or visually? Do they tend to start with one form first?
- What language do students use as they talk about steepness in the graph or changes in direction?
- Do students realize that they need to identify the graph of each swimmer's or kayaker's journey before they can answer the question?
- Do students recognize that they must think about both distance and time to find speed?
- In determining Nila's average speed in Puzzle B, several students may first find her speed during each of the four segments of her journey. Then, they may not realize that they must do more than find the average of these four speeds, as some speeds were maintained longer than others. Students need to relate the total distance to the total time. Other students may combine the times of each segment to find the total time, without recognizing that it must be 100 seconds. It is important to discuss these ideas after students have completed the puzzle.

Variations

- You can eliminate the final question on each puzzle, label each camper's story shown on the graph with a letter, and have students identify the corresponding names of the campers.
- Just provide the graph, without the story, and have students create a story about one of the parts of the graph, using their own context. Have teams exchange their stories and identify the matching representation.
- Give students the story text and have them sketch a graph to match one or more of the campers' stories.

Exit Card Choices

- What does a line segment that is parallel to the x-axis tell you?
- Pose and answer two more questions about the *Graph Story* puzzle you solved.

Extension

Invite partners to create a dramatization of a twenty-second event involving time and distance. Have each pair identify the distance to be considered and present its dramatization. Then have the "actors" show it a second time, with small groups of onlookers sketching a graph of how the distance changes over time.

Why This Game or Puzzle?

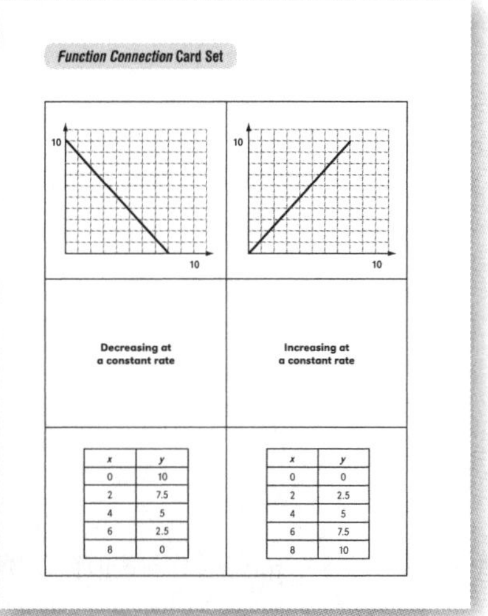

John A. Van de Walle and his coauthors suggest that graphs, equations, words, and tables can express functional relationships found in real-world contexts (Van de Walle et al. 2013). Further, translations among different representations are correlated to student success in problem solving. Students, therefore, need to recognize different representations of the same situation. In this game, students make sets of cards by finding matching graphs, tables, and descriptions.

The first three pages of cards in the set, most appropriate for sixth graders, involve change at a constant rate and include a scale, although only one number is given on each axis. The second to fourth pages are more appropriate for seventh graders, and the fifth page may be included for those students who are studying change that is non-linear, usually at the eighth-grade level.

Math Focus

- Analyzing the relationship between the independent and dependent variables from a graph, a table, or a description
- Describing qualitatively the relationship between two variables by analyzing a graph, a table, or a description

Materials Needed

- 1 *Function Connection* Card Set per group (pages A-84 and A-85, or A-85 and A-86, or A-87 and A-88)
- Optional: 1 *Function Connection* Directions per group (page A-89)

Directions

Goal: Get the most points by matching graphs, tables, and descriptions.

- Identify one player to also be the scorekeeper.
- Deal four cards to each team. Place the other cards facedown in a deck.
- Whenever a team has two or three cards in its hand that match and the other team agrees, the players put the matching set between both teams for all to see. The scorekeeper records one point for each card in the set for that team.
- Whenever a team has one card in its hand that matches a set with two cards that have already been played and the other team agrees, it adds that card to the set and the scorekeeper records one point for that team.
- Take turns being Team 1 and Team 2.
- On each turn, Team 1 begins by asking Team 2 to choose a card from its hand to show faceup. Then:
  1. If Team 1 has one or two cards in its hand that match that card, it shows the card(s) to Team 2. If Team 2 agrees, Team 1 players make a set placing all the cards together faceup, for everyone to see. The scorekeeper records one point for each card in the set.
  2. If Team 1 believes the card matches a set already made, belonging to either team, it shows the match to Team 2. If Team 2 agrees, Team 1 places the card with that set and the scorekeeper records one point.
  3. If Team 1 cannot make a match with the card, Team 2 takes the card back and Team 1 draws a card from the deck.

- When there are no cards left in the deck, play continues without the players drawing cards.
- When teams believe they have made all possible matches, the game ends, points are totaled, and the team with more points is the winner.

How It Looks in the Classroom

As *Function Connection* involves matching graphs, tables, and descriptions, this seventh-grade teacher decides to introduce the game by engaging his class in a similar task. Each student is given a card from the *Function Connection* card set and is asked to find the two other classmates who have the same situation on their cards, depicted by the two other representations.

As the students are moving around the room looking for their matched trio of cards, the teacher is recording, on the board, parts of his students' conversations. When he brings his class back together, they will discuss some of the comments made.

The teacher waits until all students think that they have found their matches, but one group of students is not certain that its three cards match. The teacher encourages the students to share the cards they have with the entire class. Gabbi comments, "I don't think that it's right, though. Shouldn't we wait until we have the right answer?" The teacher responds, "What if we all work together to see if it's right or not? That's what I'd like for you to do when you are playing the game, so this is a great opportunity to practice now."

After the class convinces this group that its cards match, the teacher directs students' attention to some of the comments written on the board, such as:

- *What does a line parallel to the* x*-axis show?*
- *What does "at a constant rate" mean?*
- *Do we have to know the exact rate when it says, "increasing at a constant rate"?*

As the teacher asks for feedback about the comments, Callie remarks, "I think that when the line is straight across, it means that there is no change in the distance." Van chimes in, "But, what if it's on the x-axis? Doesn't that mean that time is going on but there is zero of whatever the y-axis stands for during that time?" Callie responds, "Yes, that's true—same as it would be anywhere there is a horizontal line. The time moves on but the distance doesn't change."

After the class discusses the recorded comments for a few more minutes, the teacher describes how to play *Function Connection*, confident that his students experienced how to talk with each other about the math questions they might have as they play the game.

Tips from the Classroom

- It helps to have teams place their cards behind an upright file folder, allowing each team to see all of its cards easily while keeping them hidden from the opposing team.
- Some students rely too heavily on matching exact points in the table and the graph,

rather than looking for constant increases or decreases. Encourage students to look for trends rather than matching points.

- Have students describe the change they see in the graph or the table prior to comparing the card to a verbal description. You may wish to laminate the cards and allow some students to record their own descriptions on the cards showing graphs or tables.
- If appropriate, you may wish to have your students focus on the rates of change of the line segments on the graphs and compare them to the rates of change found in the tables.

What to Look For

- What connections are players making among descriptions, tables, and graphs?
- How are teams organizing their cards? For example, are they separating cards that involve an increase from cards that indicate a decrease?
- What discussions are teammates having about whether or not cards match, that you would like shared with the class?
- How are teams checking their opposing teams' matches?

Variations

- Use just two matches for each function and have students play a concentration game with the cards.
- Have students play with their cards faceup. Then on each turn, they can choose to draw a card or trade a card with their opponent.
- If your students explore the extension activity, you could make a card with a suggested real-world scenario for each set and extend play to four matching cards.

Exit Card Choices

- Write a description to match the graph below and identify the labels for the x-axis and y-axis.
- Sketch a graph to match this description: Decreasing at a constant rate, then staying constant, then increasing at an increasing rate.

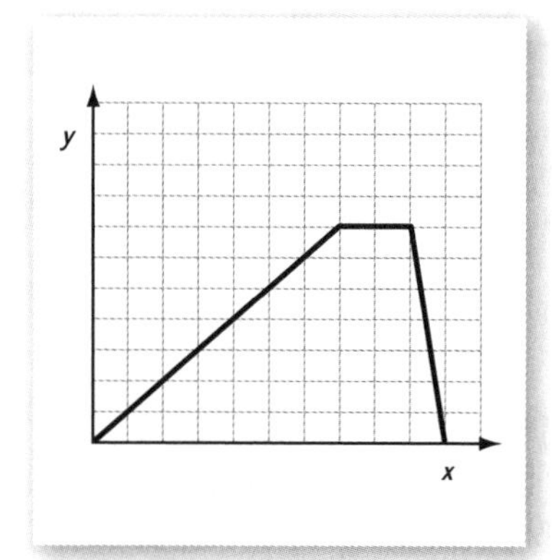

An eighth grader's response to the first exit card is shown in Figure 7.6. She demonstrated an understanding of the change, or lack of change, represented by each of the three line segments within the graph. She did not clarify the labels for the axes, though when her teacher asked about them, the student was able to identify the labels correctly. The teacher noted how much she learned about her students through their responses to this question—particularly those students who described a growing challenge they experienced, which then leveled off until it was finally mastered. This eighth grader often complained about homework, so the teacher was not surprised by the topic the student chose.

Over the 8th grade year,
I found that the amount of
homework that I had increased.
Then after the amount of homework
Stayed the same for awhile. As the
schoolyear was almost over, my
amount of homework per night decreased
to none.

Figure 7.6 Student response to the first exit card

Extension

For a week or two, show one card at the beginning of class and have students brainstorm possible matching real-world scenarios for the function.

Why This Game or Puzzle?

As eighth graders study linear functions, they learn to identify rates of change and important features on a graph, such as when the *x*- or *y*-value is zero. They also need to apply what they have learned about equations to transform an equation into its more familiar slope-intercept form. In this puzzle, students are given a variety of facts, such as: *The value of* y *is -3 when* x *is 0, The slope is 5*, or *Non-linear*. Students must match each of the facts with exactly one equation. Though the puzzle primarily provides practice with aspects of linear functions, it also requires the puzzler to reason deductively and narrow the decision space—that is, decide which facts can be immediately eliminated or identified for con-

Find One Puzzle A

Match each equation with one of its facts. You must use each fact exactly once. Write the letter of the fact under the equation.

Fact List

A. The slope is 5.	B. Non-linear	C. The value of x is 6 when y is 0.	D. The value of y is -3 when x is 0.
E. The point (5, 25) is a solution.	F. The value of y is -16 when x is 0.	G. The slope is -2.	H. The point (-2, 0) is a solution.
I. The value of x is -4 when y is 0.	J. Linear	K. The point (6, 37) is a solution.	L. The slope is -2 and the point (5, 0) is a solution.

Equation Board

$2x - 10 = -y$ ____	$y - 2 = x$ ____	$y = x^2 + 1$ ____
$-2x - 16 = y$ ____	$y = 5x$ ____	$3(x - 1) = y$ ____
$y + 6 = x$ ____	$x^2 = y$ ____	$5x + 20 = y$ ____
$5x - 3 = y$ ____	$2y + 6 = -4x$ ____	$y = -24x - 15$ ____

sideration. This problem-solving strategy, identified by George Polya in his seminal book on problem solving, *How to Solve It* (1957), is still relevant today.

Two puzzles are provided. The equations in Puzzle A are less complex, and Puzzle B includes references to y-intercepts.

Math Focus

- Deciding if a given point is a solution to an equation
- Interpreting the slope-intercept form of an equation
- Identifying equations of functions as linear or non-linear

Materials Needed

- 1 *Find One* Puzzle A or B per group (page A-90 or A-91)
- Optional: 1 *Find One* Directions per group (page A-92)

Directions

Goal: Match each equation with exactly one fact, using each fact once.

- Each fact in the list matches one or more of the equations on the equation board.
- Write each fact on a blank beneath one of its matching equations on the board. Each fact must be written exactly once.
- Write the facts so that each equation gets a match.
- Be sure to check your work.

How It Looks in the Classroom

Discussing the "do now" task on the board (Figure 7.7) allows this eighth-grade class to focus on the math involved in the puzzle they will be working on in groups during today's lesson.

Given: The linear equation: $y = 3x - 5$
True or false:

- The slope of the line is $\frac{1}{3}$.
- The point (3, 4) is a solution to the equation.
- The line crosses the y-axis above the x-axis.
- The graph of the line shows a decreasing rate.
- The y-value is 5 when $x = 0$.

Figure 7.7 "Do now" task

The teacher asks the students to first think independently and to then discuss their answers with their table groups. Each group is then assigned a statement and asked to present to the class why it chose true or false. Rafi's group has not yet reached an agreement about its answer to the last statement, so the group decides to present its dilemma to the class. He explains, "Most of us think that the answer is true, but Alexia thinks that it's false. We think that it's true because there is a 5 on the end of the equation. Alexia says that it's a negative 5. We're not sure how to figure out which is right." As students offer their suggestions for how to reach a solution, Joshua points out that there is another way to find the answer. "We looked more at the $x = 0$ statement. We substituted 0 for x and solved for y, and found that $y = -5$. So we think that statement is false." Rafi remarks that he now agrees with Alexia's answer and likes the strategy that Joshua had to offer.

The teacher asks the students to consider each of the true/false statements again, this time finding a different way than they found before to explain why they think that a statement is true or false. The connection between their work in the introductory activity and in the *Find One* puzzle will become more apparent to the students as they work on the puzzle in their small groups.

Tips from the Classroom

- In our field-testing, most students began by making random choices and then realized that their choices were not going to allow them to match each fact with exactly one equation. As they persevered, they began to make lists or to recognize facts with fewer matching equations (such as *Non-linear*).
- A few students became impatient with the number of times they had to erase. It helped them to list all the choices for each equation in ink and then cross out choices in pencil. Others preferred to cut apart the facts and move them around as they made choices.

What to Look For

- What problem-solving strategies do students use? Do they make lists? Do they use guess and check?
- How are students using the process of elimination? Do they immediately recognize the two non-linear equations in Puzzle A? What other ways are puzzlers finding to be efficient?
- What partial understandings do you observe? For example, in Puzzle B, do students understand that if the y variables have coefficients other than one, they'll need to do some computation before identifying the slope?
- What strategies or conversations do you want shared with the class when you debrief?

Variations

- This puzzle could also be a game. Cut apart the facts to use as cards. To play the game, students deal all the cards, and on each turn a team places a fact on one of the equations on the shared equation board. Only one card may be placed on each equation. Teams alternate turns. The game ends when a player cannot place a card. The team with the fewest cards left in its hand wins.

- You can replace the *y*-intercept language in Puzzle B with *The value of* y *is … when* x *is 0*. Or, you could incorporate x-*intercept* to replace facts that read *The value of* x *is … when* y *is 0*.

Exit Card Choices

- If the slope of a line is 4 and it contains the point (3, 6), what is the equation of the line? Show why it is the correct equation.
- Which fact was the most helpful to you in the beginning? Why? Which fact was the least helpful to you in the beginning? Why?

Extension

Give each student a fact card and post equations around the classroom. Have each student write his or her name next to any equation the fact card matches. Show students a list of all the facts and have them try to identify, based on where names are written, who had which fact.

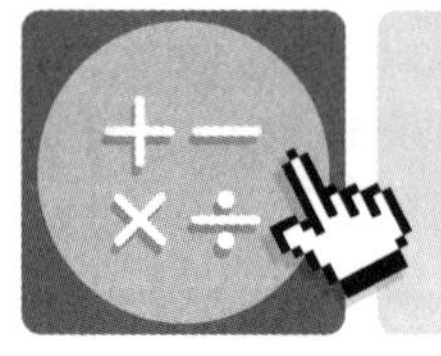

Online Games and Apps

There are a myriad of interesting online games and puzzles in which students can explore the concept of patterns. Visual images, number patterns, and graphs are all easily depicted using technology, often allowing the user to manipulate the data provided. Determining the rate of change or the accuracy of a chosen output to match an input, or visualizing the graph of an equation, are each flexibly managed as a result of the power of technology. Several examples of online games that demonstrate these characteristics include:

- Stop That Creature (free from http://www.learner.org/teacherslab/math/patterns/mystery) engages players in a goal to stop clones from destroying Poddleville. Inputs are sent into a function machine, which then produces the corresponding outputs. Each time the player finds the right rule, as an algebraic expression, a clone is stopped. An input-output table, labeled *before* and *after*, is built prior to the player guessing the rule. Players who choose an inaccurate expression as the rule are given the opportunity to try again, and hints are provided if players choose to use them. Players must stop ten clones to win the game.

- Function Carnival (free from http://teacher.desmos.com) allows users to make connections between what is seen in a short video—for example, a parachutist shot out of a cannon—and a graph showing a change in one variable over time. Once students draw their own graph of what they see in the video, they are able to compare the actual graph of what happens in the video with their own graph. Each of the three scenarios increases the difficulty level of the graphs. Teachers who have chosen to populate a class on the website can view and save their students' graphs for class discussions.

- Save the Zogs (free from http://mathplayground.com) engages players in analyzing or creating equations to rescue the Zogs, four aliens who are hidden in a straight line on a coordinate graph. If players create an accurate equation to represent the line the Zogs are on, they rescue them. Different levels allow players to manipulate characteristics of the equations based first on horizontal and vertical lines, then on lines with a slope other than zero or undefined. The ability to set a slope and *x*- or *y*-intercept, and then to graph the equation, gives players immediate feedback about the choices made, allowing them to learn from each play session.

APPENDIX A

Etiquette Expert Cards

Dear Etiquette Expert, One of the students in my class always wants to be the one to say what we should do when it is his team's turn. He doesn't give his partner a chance, no matter who it is. We feel like we understand the math, but are just not as quick as he is. How can we help him to understand that his partner needs to be involved in deciding what to do as well?	Dear Etiquette Expert, I am a very shy person and get nervous when we have to find our own partners. Sometimes I go sharpen a pencil or something and feel silly that I don't know how to find a partner myself. Often I wait until my teacher notices that I don't have a partner and she assigns me to someone. Can you help me figure out how I can do this more on my own?
Dear Etiquette Expert, My partner and I were working on a puzzle, but after working on it for a little while she just gave up. She said it was too hard and we should do something else instead. I thought we could solve it if we worked together. How do I convince her to work longer and that the effort is worth it?	Dear Etiquette Expert, My team won the game we were playing today, but my partner made a big deal about our win and I felt really embarrassed. He kept telling the other team that we were much better than they were and that we would win the next time that we played, too. How can I avoid this big scene again?
Dear Etiquette Expert, Sometimes I really like solving problems alone, but my teacher likes team members to work together. I get too distracted when my partner starts talking right away and can't think about the math. I want to ask if we can start a turn by reading and thinking to ourselves, but I don't want to seem too different from everyone else. What do you think?	Dear Etiquette Expert, My partner and I had a lot of questions about the puzzle today. We read the directions again, but it didn't help and everyone else looked like they knew what they were doing. We finally just filled in numbers randomly and said we were done. I know we need to find a way to work through being stuck. Can you help us figure out how to keep going?

Match It Puzzle A Pieces

<table>
<tr><td>250%
0.5%</td><td>2.5
0.6
0.3</td><td>60%
$\frac{3}{100}$
0.03</td><td>0.03
7.5%</td></tr>
<tr><td>0.005
1.5
75%</td><td>$\frac{3}{10}$
150%
3.4
50%</td><td>3%
340%
0.7
$\frac{1}{5}$</td><td>0.075
70%
25%</td></tr>
<tr><td>$\frac{3}{4}$
0.02
0.4</td><td>$\frac{1}{2}$
2%
$\frac{1}{10}$
4%</td><td>0.2
0.1
$\frac{2}{5}$
5%</td><td>0.25
40%
7.5</td></tr>
<tr><td>$\frac{2}{5}$
20%</td><td>$\frac{1}{25}$
$\frac{1}{5}$
0.01</td><td>$\frac{1}{20}$
$\frac{1}{100}$
2%</td><td>750%
$\frac{1}{50}$</td></tr>
</table>

Match It Puzzle B Pieces

<table>
<tr>
<td>Greater than 100%
0.5%</td>
<td>2
10% greater than $\frac{1}{2}$
0.3</td>
<td>60%
$\frac{3}{100}$
0.03</td>
<td>Less than 5%
7.5%</td>
</tr>
<tr>
<td>0.005
1.5
Equal to 75%</td>
<td>Greater than $\frac{1}{4}$
150%
3.4
Less than 30%</td>
<td>3%
$\frac{34}{10}$
0.7
Less than $\frac{1}{4}$</td>
<td>Between 0 and 0.1
Greater than 50%
25%</td>
</tr>
<tr>
<td>$\frac{3}{4}$
0.02
Between 55% and 65%</td>
<td>$\frac{1}{4}$
2%
$\frac{9}{10}$
4%</td>
<td>0.2
10% less than 1 whole
$\frac{4}{5}$
5%</td>
<td>0.25
Greater than 75%
700%</td>
</tr>
<tr>
<td>0.6
Less than 0.4</td>
<td>$\frac{1}{25}$
$\frac{1}{3}$
0.05</td>
<td>$\frac{1}{20}$
Less than 10%
4%</td>
<td>Greater than 5
$\frac{2}{50}$</td>
</tr>
</table>

Match It Puzzle C Pieces

7^2 0.003	49 5^3 64	125 1.01×10^1 10	10.1 9×10^3
3×10^{-3} 60 2500	8^2 6×10 3×10 0.009	10^1 30 125 4^2	9000 1.25×10^2 3^3
2.5×10^3 30^2 0.4	9×10^{-3} 900 2.5×10^{-1} 1^2	16 0.25 1000 0.64	27 10^3 6^2
$\frac{2}{5}$ 6.4×10^1	1 64 0.00101	6.4×10^{-1} 1.01×10^{-3} 2.5×10^2	36 250

Make Your Own *Match It* Puzzle Template

Match It Directions

Directions

Goal: Arrange each puzzle piece so that the numbers represented on all adjoining sides match.

- Work together.
- Organize each of the 16 puzzle pieces into a square.
- The pieces should be placed so that the numbers on each of their sides match; that is, the representations on the sides of any adjacent squares have the same value.
- Check to make sure you have matched each side correctly.

Tic-Tic-Tac-Toe Game Board A

Dividend

315	7,875
18,900	
2,835	1,260

Divisor

9	35
105	
15	3

3	140	6,300	81	525
84	180	315	2,625	9
540	189	75	35	420
945	225	21	12	2,100
875	105	36	1,260	27

Tic-Tic-Tac-Toe Game Board B

Dividend	Divisor
$1\frac{3}{4}$, 6, $3\frac{1}{2}$, 2, $3\frac{1}{3}$	$\frac{3}{8}$, 3, $1\frac{1}{4}$, 4, $\frac{5}{6}$

$4\frac{2}{3}$	2	$2\frac{4}{5}$	$\frac{1}{2}$	4
$4\frac{4}{5}$	$\frac{7}{8}$	$2\frac{2}{5}$	$8\frac{8}{9}$	$\frac{7}{12}$
$4\frac{1}{5}$	$5\frac{1}{3}$	$1\frac{1}{9}$	$1\frac{2}{5}$	$1\frac{1}{2}$
$\frac{2}{3}$	$2\frac{2}{3}$	$\frac{7}{16}$	$7\frac{1}{5}$	$9\frac{1}{3}$
$\frac{5}{6}$	$2\frac{1}{10}$	16	$1\frac{1}{6}$	$1\frac{3}{5}$

Tic-Tic-Tac-Toe Game Board C

Dividend

360		0.2
	6.3	
40		824

Divisor

3		0.1
	0.8	
0.4		6

$0.0\overline{6}$	120	2	7.875	100
$13.\overline{3}$	8,240	450	0.5	1.05
900	$0.0\overline{3}$	2.1	400	1,030
0.25	15.75	$6.\overline{6}$	$274.\overline{6}$	3,600
63	50	2,060	60	$137.\overline{3}$

Tic-Tic-Tac-Toe Directions

Directions

Goal: Choose divisors and dividends in order to mark off four quotients in a row, column, or diagonal on the game board.

- Decide which team will be *X* and which will be *O*. The first team picks one number from the set of dividends and one from the set of divisors. Both teams compute the quotient. (Teams get to pick only once, even if they discover that they don't get a quotient they want.)
- Once both teams have confirmed the quotient, the first team finds it on the game board and writes the team's mark (*X* or *O*) in that space.
- If the team gets a quotient that is already marked with an *X* or an *O*, it loses its turn.
- Teams alternate turns.
- The first team to write its mark in four touching quotients in a row, column, or diagonal is the winner.

Three-For Game Board

-1	7	22	-4	5	-25
1	-18	9	20	18	6
10	-19	-7	8	-12	-16
-14	25	11	-15	2	-8
-10	14	-9	19	-16	3
-21	0	-11	6	-3	-20

Three-For Recording Sheet

Students on Team A ______________________________

Students on Team B ______________________________

Target Number	Num-bers Used	Equation	Team A Points	Team B Points
			Total Points	**Total Points**

Three-For Directions

Directions

Goal: Create sets of three cards each that equate to a target number.

- Give each team a copy of the game board.
- One team draws a card from the deck and places it for all to see. If the card is red, it is a negative integer; if the card is black, it is a positive integer.
- Teams are to find three numbers on the board that can be used in an equation (all four operations are allowed) to equal the target number.
- When they find the three numbers, they call "Three-For" and the other team must stop looking for a solution. They should record their equation on the recording sheet.
- The equation is checked by the opposing team. If it is correct, the team that found the three numbers gets 1 point and crosses the three numbers off the board. Additional scoring will occur if the team finds three numbers in which two are adjacent, allowing two points to be scored, and three points if all three are adjacent.
- If the equation is incorrect, both teams return to finding a solution.
- Play continues until there are no more numbers to choose from or there are no more possible "*Three-Fors*" when a target number is displayed.

Communicate It! Cards

INTEGER	RATIONAL	IRRATIONAL	REAL
SQUARE NUMBER	CUBIC NUMBER	SQUARE ROOT	CUBE ROOT
WHOLE NUMBER	NATURAL	POSITIVE	NEGATIVE
SCIENTIFIC NOTATION	ABSOLUTE VALUE	DISTRIBUTIVE PROPERTY	COMMUTATIVE PROPERTY
ASSOCIATIVE PROPERTY	IDENTITY PROPERTY	MULTIPLICATIVE INVERSE	PRIME FACTORIZATION

Communicate It! Directions

Directions

Goal: Guess the vocabulary word.

- Each team chooses a card from the deck, silently reads the vocabulary word, and plans privately how they will communicate the term.
- The timer is set for 2 minutes.
- Team 1 uses words or actions to communicate its vocabulary word to the other team, without saying the word itself.
- If Team 2 is able to guess the vocabulary word before the timer runs out, the teams get 1 point.
- The other team then tries to communicate the word on its card.
- Play continues until each team has had the opportunity to try to communicate four vocabulary terms.
- If, together, the teams earn 6 points, they win the game.

A-Maz-ing! Game Board

START	-4	$\frac{1}{8}$	0	$\sqrt{25}$	-1.2
3.7	$\sqrt{81}$	1.6	7^2	0.4	$\sqrt{1}$
-3	0.7	$\frac{2}{5}$	-6	$\sqrt{16}$	1.4
$\frac{1}{4}$	5	-1.3	4^2	2.5	$\frac{5}{6}$
5.8	5^2	1.2	$\frac{1}{3}$	16	6
10	$\frac{3}{8}$	$(-4)^2$	-5	6^2	0.8
-3^2	3	-1^2	0.3	-0.4	$\frac{4}{5}$
4	-10	4	$\frac{5}{8}$	$(-2)^2$	**FINISH**

A-Maz-ing! Recording Sheet

Student(s) ______________________________

Turn	Numbers Used	Equation	Total
1			
2			
3			
4			
5			
6			
7			
8			
9			

Game Total ______________

A-Maz-ing! Directions

Directions

Goal: Make your way through the maze by creating equations, ending with the least (or greatest) possible total.

- Each team places a chip on the START.
- Teams take turns rolling the die and moving that number of places.
- Players may move their chip to the right or down (or any combination of these two directions), but not left or up.
- After a team has moved its chip the correct number of places, it finds the total of all of the numbers that the chip passed over or landed on while moving through the maze. All operations may be used, and numbers may indicate exponents or be included in scientific notation.
- The team then writes the equation on the recording sheet.
- When one team makes it to the FINISH, the totals are added. The team with the least (or greatest) total wins the game.

Four of a Kind Card Set A

$6\frac{1}{2}$ to 4	13 39 117 8 24 72
$\frac{26}{16}$	
5 for every 3	**15 to 9**
	10 20 30 6 12 18
20 30 40 24 36 48	$\frac{10}{12}$
	15:18
$\frac{4}{5}$	
36 per 45	8 12 16 10 15 20

Four of a Kind Card Set B

$\frac{10}{34}$

25 to 85

20	30	40
68	102	136

20 for every 24

$\frac{80}{96}$

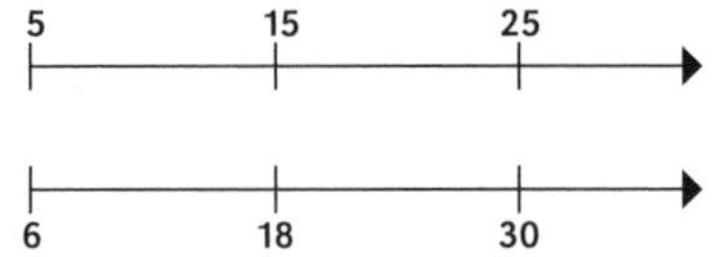

5 15 25

6 18 30

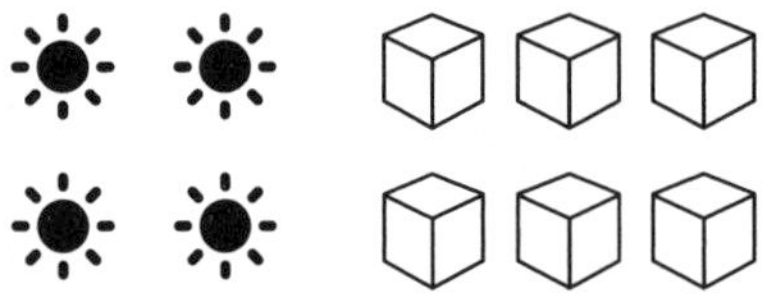

$\frac{14}{21}$

2 to 3

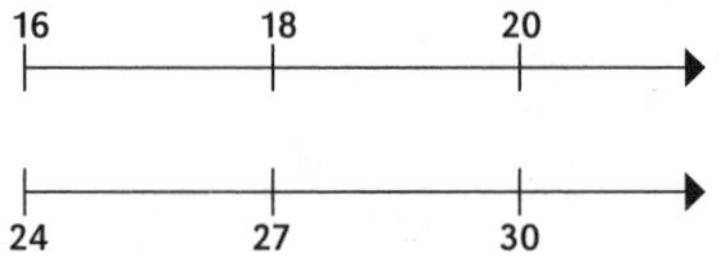

16 18 20

24 27 30

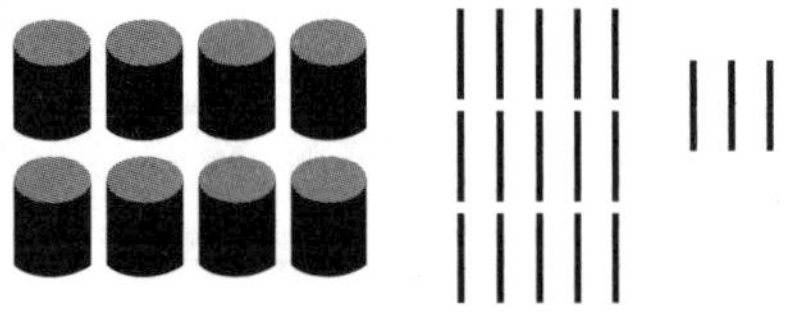

4 for each 9

28	32	36
63	72	81

$\frac{24}{54}$

Four of a Kind Directions

Directions

Goal: Be the first to collect four playing cards with equivalent ratios.

- Shuffle the cards.
- Deal four cards to each player.
- At the same time, each player looks at his or her own cards and decides on one card the player does not want. The player places that card facedown in front of the player to the right.
- All the players pick up their new cards, at the same time, so that each person once again has four cards in his or her hand.
- Players continue to pass and pick up cards, waiting for all players to pick up before the next pass begins.
- The first player to get four cards with equivalent ratios says, "Four of a Kind!" and wins.

Best Unit Price Cards

Marie's Moisturizing Shampoo 1 pt. 2 oz. $9.00	Marie's Moisturizing Shampoo 20 fl. oz. $11.00	Marie's Moisturizing Shampoo 24 fl. oz. $15.50	Marie's Moisturizing Shampoo 12 fl. oz. $8.00
Fine Line Black Pens Box of 100 $89	Fine Line Black Pens Pack of 10 $12.99	Fine Line Black Pens Pack of 5 $7.50	Fine Line Black Pens Pack of 2 $3.50
$\frac{3}{8}$″ -wide Ribbon 12 yards $3.25	$\frac{3}{8}$″ -wide Ribbon 27 feet $2.75	$\frac{3}{8}$″ -wide Ribbon 7 yards $2.50	$\frac{3}{8}$″ -wide Ribbon 20 feet $2.50
Seedless Green Grapes 1 lb. 6 oz. $3.25	Seedless Green Grapes 2 lb. 10 oz. $7.00	Seedless Green Grapes 24 oz. $4.50	Seedless Green Grapes 18 oz. $3.60

Best Unit Price Cards *(continued)*

Cattails Cat Food 15 cans $7.50	Cattails Cat Food 18 cans $9.50	Cattails Cat Food 12 cans $6.75	Cattails Cat Food 20 cans $11.50
Bea's Beads 100 beads/bag $4.95	Bea's Beads 150 beads/bag $7.50	Bea's Beads 25 beads/bag $1.50	Bea's Beads 50 beads/bag $3.50
Sun-Ease Sunscreen 22 fl. oz. $11.00	Sun-Ease Sunscreen 10 fl. oz. $6.50	Sun-Ease Sunscreen 12 fl. oz. $8.00	Sun-Ease Sunscreen 8 fl. oz. $6.00
Twist-to-Write Pencils Pack of 10 $7.50	Twist-to-Write Pencils Box of 75 $70.00	Twist-to-Write Pencils Pack of 5 $4.95	Twist-to-Write Pencils Pack of 3 $3.50

Best Unit Price* Cards *(continued)

Smelly Stickers 10 sheets/pkg $10.25/pkg	Smelly Stickers 5 sheets/pkg $6.50/pkg	Smelly Stickers 3 sheets/pkg $4.50/pkg	Smelly Stickers single sheet $1.95
Wonder Water 24 bottles/case $7.99	Wonder Water 36 bottles/case $15.00	Wonder Water 6 bottles/pack $3.00	Wonder Water 12 bottles/pack $8.00
Oliver's Olive Oil 28 fl. oz. $27.95	Oliver's Olive Oil 2 pints $34.95	Oliver's Olive Oil 12 fl. oz. $15.95	Oliver's Olive Oil $\frac{1}{2}$ pint $34.95
Magic Carpet 4 ft. x 6 ft. $59.95	Magic Carpet 48 in. x 84 in. $84.00	Magic Carpet 36 in. x 48 in. $40.00	Magic Carpet 6 ft. x 5 ft. $120.00

Best Unit Price Cards *(continued)*

Directions

Goal: Collect the most cards.

- Shuffle the cards and deal each team six cards, which the teams keep private.
- Place the remaining cards facedown as a deck.
- On each turn:

1. Turn over the top card of the deck.
2. Any team that has a card in its hand with the same product as the card turned over puts that card faceup in front of its players.
3. Players compare to see which card shows the best unit price. If a team has the card, it puts these cards in its pile of winning sets. If the deck's card shows the best unit price, all the played cards are placed facedown in a discard pile.
4. Teams take a card from the top of the deck to replace any cards played, keeping six cards in their hand as long as possible.
5. If neither team has a card with the same product as the deck's card, that card is placed in the discard pile.

- When all of the cards in the deck have been turned over once, teams take turns putting down a card. If the other team has a card with the same product, the unit prices are compared. If the other team does not, the card is added to the discard pile.
- When teams no longer have cards with matching products, the cards in the winning sets are counted and the team with the greater number of cards wins.

Keep Going? Cards

Skateboard 40 minutes 7.5 mph	Rollerblade 20 minutes 10 mph	Walk 40 minutes 3.5 mph	Run 30 minutes 6 mph
Jog 75 minutes 4 mph	Bike 20 minutes 12 mph	Unicycle 15 minutes 2 mph	Car 5 minutes 60 mph
Skateboard 30 minutes 12 mph	Rollerblade 40 minutes 10 mph	Walk 60 minutes 4 mph	Run 30 minutes 5 mph
Jog 45 minutes 4 mph	Bike 24 minutes 15 mph	Unicycle 30 minutes 3 mph	Car 10 minutes 42 mph

Keep Going? Recording Sheet

Name(s): ______________________________ **Date:** ______________

Card 1:

Miles for this part of the trip: __________

Total miles so far: ______________

Card 2:

Miles for this part of the trip: __________

Total miles so far: ______________

Card 3:

Miles for this part of the trip: __________

Total miles so far: ______________

Card 4:

Miles for this part of the trip: __________

Total miles so far: ______________

Card 5:

Miles for this part of the trip: __________

Total miles so far: ______________

Card 6:

Miles for this part of the trip: __________

Total miles so far: ______________

Keep Going? Directions

Directions

Goal: Get a total distance closest to 21 miles without going over.

- Choose a player to also serve as the dealer. The dealer shuffles the cards and deals two cards to each team with one card faceup and one card facedown.
- Teams may look at the facedown card, but keep it private from the other team.
- Place the remaining cards facedown as a deck.
- On each turn:
  1. Each team may choose to "stop" (not take any more cards) or to "keep going" (request another card from the dealer).
  2. If a team chooses to "keep going" by receiving another card, the dealer turns over two cards from the deck and the team chooses one of the cards. The card not chosen is returned to the bottom of the deck.
  3. The team determines how close it is to 21 miles and records its mileage on the recording sheet.
  4. The team continues steps 2–3 until it chooses to stop.
- Once both teams have had their turn, they each reveal how close they are to 21 miles.
- The team that is the closest to 21 miles, without going farther, is the winner.
- A new dealer is chosen for the next game and all cards are then returned to the deck and shuffled, and play begins again.
- Play continues for a predetermined period of time.

The Question Is/The Answer Is Cards

The Numerical Answer Is 112 *The Question Is* Kai's birthday party will cost $120 if he invites 10 guests. If 2 more guests come to the party, how much will it cost for Kai's party?	*The Numerical Answer Is* 144 *The Question Is* If it costs you $55.80 to fill your 20-gallon gas tank, but you only have $34, how much gas (to the nearest gallon) could you buy?
The Numerical Answer Is 144 *The Question Is* A store has T-shirts on sale at 3 for $16.50. If a customer spent $66 on T-shirts, how many did he buy?	*The Numerical Answer Is* 12 *The Question Is* The Garcia family traveled 273 miles in 4.2 hours on the first part of their vacation. If they traveled at the same rate to get to their next destination, how far would they travel in 6 hours?
The Numerical Answer Is 12 *The Question Is* You invest $3,600 in a stock, for one year, that pays a $162 dividend. At the same rate, how much will you need to invest to earn $270?	*The Numerical Answer Is* 390 *The Question Is* The student council representatives determine that they can raise $85.80 if 13 people buy water bottles. How much money will they raise if 72 people buy water bottles?
The Numerical Answer Is 6,000 *The Question Is* Marcus works 15 hours mowing lawns and makes $216. How much money will he make if he works 10 hours?	*The Numerical Answer Is* 475.20 *The Question Is* If an airplane can travel 2,400 miles in 6 hours, how far can it travel in 15 hours?

The Question Is/The Answer Is Cards *(continued)*

The Numerical Answer Is 6,000 *The Question Is* The scale factor from a replica to an original tall ship is 1:84. If the height of the tall ship is 218.4 feet, how tall will the replica be?	*The Numerical Answer Is* 5 *The Question Is* Chris can walk 0.7 miles in 12 minutes. At this rate, how many miles can he walk in an hour?
The Numerical Answer Is 2.6 *The Question Is* Jon and Erica took turns shoveling for 10 hours and split the money they earned according to how many hours they worked. If they earned a total of $200, and Erica worked for 6 hours, how much did Jon earn?	*The Numerical Answer Is* 3.5 *The Question Is* Annika is starting a new business and is marketing by sending e-mails to potential customers. If she has time to send 58 e-mails in 2 days, how many e-mails can she send in 5 days?
The Numerical Answer Is 80 *The Question Is* Sammie figured out that she drove 160 km, which was the same as 99 miles (to the nearest mile). If she finished her trip by driving another 221 km, how many total miles did she drive (to the nearest mile)?	*The Numerical Answer Is* 145 *The Question Is* Tham took care of his neighbors' dog for 12 weeks while they were away and earned a total of $944. If he was paid every three weeks, how much did he receive for each payment?
The Numerical Answer Is 236 *The Question Is* During a blizzard, snow was falling at a rate of 4.2 inches/hour. If it kept falling at the same rate, how many hours would it take for 1.75 feet to fall?	*The Numerical Answer Is* 236 *The Question Is* I spent $70 on 20 yards of fabric for the scenery for the school play. If I have to buy another 12 yards of the same fabric, how much money will I spend in all?

The Question Is/The Answer Is Make Your Own Puzzle Cards

The Numerical Answer Is ____________________ *The Question Is*	*The Numerical Answer Is* ____________________ *The Question Is*
The Numerical Answer Is ____________________ *The Question Is*	*The Numerical Answer Is* ____________________ *The Question Is*
The Numerical Answer Is ____________________ *The Question Is*	*The Numerical Answer Is* ____________________ *The Question Is*
The Numerical Answer Is ____________________ *The Question Is*	*The Numerical Answer Is* ____________________ *The Question Is*

The Question Is/The Answer Is Directions

Directions

Goal: Place cards so that the answer identified on each card answers the question on the card before it.

- Spread out the cards faceup on the table or the floor.
- Choose a card and read its question.
- Find a card with a matching answer, place this card next to the first card, and read the question on this second card.
- Continue to read questions and find answers. Organize the cards in a circle so that each question is followed with a correct answer.
- Each card must be included in the circle.

Money Matters Puzzle A

Name(s): ______________________________ **Date:** ______________

Lewis, Marie, Norah, Odette, and Portia each went shopping. They each spent a different amount of money and a different percentage of the money they brought shopping. The amounts of money they spent were $25, $36, $54, $70, and $100. They spent 12.5%, 25%, 50%, 75%, and 90% of their shopping money. Use these clues to find how much money each person had for shopping.

- Lewis had $80 more shopping money than Norah.
- Norah and Portia both had the same amount of shopping money.
- Marie had less than $50 of shopping money.
- Odette spent three-fourths of her shopping money.
- Portia spent four times as much as Norah spent.

	Amount of Money Spent					Percent of Money Spent				
Names	$25	$36	$54	$70	$100	12.5%	25%	50%	75%	90%
Lewis										
Marie										
Norah										
Odette										
Portia										

Lewis had $__________ for shopping.

Marie had $__________ for shopping.

Norah had $__________ for shopping.

Odette had $__________ for shopping.

Portia had $__________ for shopping.

Money Matters Puzzle B

Name(s): ______________________________ **Date:** ______________

Andy, Bwan, Carry, Dansi, and Ethan work during the summer. They babysit, mow lawns, and walk dogs. This summer they each made more money than last summer.
The rates of increase were: 1%, 3%, 5%, 8%, and 10%.
The total amounts they earned this summer were: $515, $601, $693, $726, and $999.
Use these clues to find how much money each person earned last summer.

- Each person earned a different amount this summer and had a different rate of increase.
- Ethan's rate of increase was less than Bwan's, but he earned more money last summer than she did.
- Dansi and Andy were surprised that they both earned the same amount last summer.
- The total amount Carry earned last summer was between $910 and $940.
- Dansi's rate of increase was more than Andy's.

	Amount Earned This Summer					Rate of Increase				
Names	$515	$601	$693	$726	$999	1%	3%	5%	8%	10%
Andy										
Bwan										
Carry										
Dansi										
Ethan										

Andy earned $__________ last summer.

Bwan earned $__________ last summer.

Carry earned $__________ last summer.

Dansi earned $__________ last summer.

Ethan earned $__________ last summer.

Money Matters Directions

Directions

Goal: Use clues to match the right pieces of information (for example, a person's name with the money he or she spent and a person's name with the total amount of money he or she borrowed).

- Work alone or with a partner.
- Read the clues.
- Use the table to organize what you know from each clue.
- Make notes so that you can recall your thinking. Include computation and clue numbers in your notes so that you can refer back to them if you need to.
- You do not need to use the clues in order.
- When you are finished, check your solution with each clue to be sure that it works.

Find It Together Puzzle Cards A

On Wednesday the 15th of June, Mayara bought 6 pencils.	On Sunday the 19th of June, she had 20 pencils left.
On Saturday the 18th of June, she found 5 more pencils in her drawer.	On Friday the 17th of June, she gave half of her pencils to her soccer teammates.
On Thursday the 16th of June, she bought three times the amount she had on Tuesday the 14th of June.	How many pencils did Mayara have on Tuesday the 14th of June? Record the equation determined by the group.

Find It Together Puzzle Cards B

Later in the week, he paid $5 for a ticket to the school concert.	Giovanni was given his allowance at the beginning of the week.
He earned an amount equivalent to his allowance by raking leaves at his neighbor's house.	He spent half of his allowance on dinner with his friends.
He earned $7 by delivering newspapers.	He ended the week with $29. What was his weekly allowance? Record the equation determined by the group.

Find It Together Directions

Directions

Goal: Use the clues to find the mystery equation and solution.

- Form a team of three puzzle solvers.
- Place the clues facedown. Each solver randomly takes two of the clues.
- Solvers may read their clues to the others, but may not show the clues.
- Work together to figure out the equation and then the solution being described by the clues. Read the clues as many times as necessary, and talk about what you know.
- You can write or draw to help you understand the information in the clues, but you can't record the clues.
- When you think that you have the equation and the solution, read the clues again to check.

Equivalent Expressions Cards

$-(x - 12)$	$12 - x$
$2x - 2 + 5$	$3 + 2x$
$2x - (-2) + 24$	$2(x + 13)$
$7x + 5x - 2$	$-2 + x(7 + 5)$
$2x - 7$	$-3 - 4 - (-2x)$
$-3(x - 7)$	$30 - 3(3 + x)$
$20 - 3x - 1 + 2$	$-3x + 21$
$12x + 15y$	$3(4x + 5y)$
$15y + 12x$	$-3(-4x + (-5y))$

Equivalent Expressions Cards *(continued)*

$24x + 72y$	$(2x + 6y)12$
$9x + 3 - 2x - 15$	$7x - 12$
$-(7x + 12)$	$-12 - 7x$
$48 - 3x + 4(x - 9)$	$x + 12$
$\frac{(35x + 45)}{5}$	$7x + 9$
$\frac{(24x - 64)}{4}$	$2(3x - 8)$
$-12x - 2(-9x + 4)$	$-8 + 6x$
$-16x + 12$	$12 - 16x$
$4(3 - 4x)$	$-(22x + 8 - 6x + (-20))$

Equivalent Expressions Cards *(continued)*

$\frac{3}{4}x + 1\frac{1}{2}$	$\frac{3}{4}(x + 2)$
$\frac{-(3 + 6x)}{0.5}$	$-6(2x + 1)$
$0.4(-10x + 5)$	$2 - 4x$
$19 - 0.2(10x + 10)$	$17 - 2x$
$9 - 0.34x + 2x - 0.66x$	$x + 9$
$-0.75x + 1 + 0.5x - 0.5$	$-0.25x + 0.5$
$-0.5(0.5x - 1)$	$0.75x + 0.5 + (-x)$
$\frac{-(\frac{1}{4} + \frac{3}{4}x)}{\frac{1}{4}}$	$-(1 + 3x)$
$\frac{-(\frac{3}{2}x + \frac{1}{2})}{\frac{1}{2}}$	$7.5x - 6 - 10.5x + 5$

Equivalent Expressions Recording Sheet

Name(s): ______________________________ **Date:** ______________

______________________ = ______________________

______________________ = ______________________

______________________ = ______________________

______________________ = ______________________

______________________ = ______________________

______________________ = ______________________

______________________ = ______________________

______________________ = ______________________

______________________ = ______________________

______________________ = ______________________

Equivalent Expressions Directions

Directions

Goal: Collect more pairs of equivalent cards (packs).

- Shuffle the cards. Deal each team four cards faceup for all to see. Put the other cards facedown in a pile.
- On each turn you can do one of three things:

  1. Find two of your cards that have equivalent expressions. Set this pair beside you. Replace them with two cards from the top of the pile.
  2. Trade one of your cards with one of the other team's cards when you are able to make a pair by doing so. Set this pack beside you. Each team replaces its card with a card from the top of the pile.
  3. Draw a card from the top of the pile and add it to your cards.

- When a pair is made, both teams must agree that the expressions are equivalent and then one player records their thinking on the recording sheet.
- If no cards are left in the pile, you can still have a turn, but you don't replace any cards you play.
- The game ends when no team can make another pair.
- The team with more pairs wins.

Solve It Game Board

Equations: | **Solutions:**

$\_\_\_\_\_ x + \_\_\_\_\_ = \_\_\_\_\_ x - \_\_\_\_\_$ | $x = \_\_\_\_\_$

$\_\_\_\_\_ - \_\_\_\_\_ x = \_\_\_\_\_$ | $x = \_\_\_\_\_$

$\_\_\_\_\_ (\_\_\_\_\_ x - \_\_\_\_\_) = \_\_\_\_\_$ | $x = \_\_\_\_\_$

$\_\_\_\_\_ (\_\_\_\_\_ + \_\_\_\_\_ x) = \_\_\_\_\_ x$ | $x = \_\_\_\_\_$

$\_\_\_\_\_ x + \_\_\_\_\_ = \_\_\_\_\_ - \_\_\_\_\_ x$ | $x = \_\_\_\_\_$

Discard ______ **Sum of the values for** $x =$ ______

Solve It Directions

Directions

Goal: To score either the lesser (or greater) number of points as a result of solving equations.

- Decide if the lesser or greater number of points wins.
- Shuffle the cards and place them facedown. Choose a player to also be the game leader.
- The game leader turns over one card and announces the number. Each team writes the number in one of the spaces on the game board. A discard box is provided for one of the numbers.
- Both teams must record the number before the next number is announced and, once written, its placement cannot be changed.
- The game leader continues turning over cards until twenty numbers have been recorded.
- Both teams then solve the equations that were created by recording numbers in the spaces.
- Each team finds the total of its x-values. If an equation results in a value for x that is not possible, the team adds 0 points to the total. For each equation that results in a value for x for which the solution is all real numbers, the team scores -5 (or 5) points. The team with the lesser (or greater) sum is the winner.

Express It! Written Expressions Card Set A

1 Thirty-two divided by the quantity two times a number plus four	2 The difference of three times a number and ten
3 The square of a number increased by six	4 Six less than the product of two and a number
5 Three times the quantity of three plus a number	6 The sum of twice a number and five times the same number
7 The quotient of seven squared divided by a number	8 The quantity of eight times a number plus two, cubed
9 Four times a number decreased by eleven	10 Eight divided by four times a number

Express It! Written Expressions Card Set B

1 A painter charges \$30 per hour and spends \$10 a day on gasoline. Write an algebraic expression to represent his earnings for one day.	2 Alex made three withdrawals of the same amount (*w*) from his savings account, which originally had a balance of \$400. Write an algebraic expression to represent the amount he has after the withdrawals.
3 Merrick earns \$95 a day working part time at a supermarket. He already has \$500 saved for his college spending money. Write an algebraic expression to represent the amount of money he will have for spending money in (*d*) days.	4 Marguerite is twenty years old. Write an algebraic expression to represent how old Marguerite will be in (*y*) years.
5 One hundred students will share some pizzas (*p*) that have each been sliced into 8 pieces. Write an algebraic expression to represent how many pizza slices each student will get.	6 Alishia had \$55.25 to spend at the mall. She bought a pair of pants for \$25.95 and 5 T-shirts (*t*). Write an algebraic expression to represent the amount that Alishia had left after her spending spree at the mall.
7 Terrence has (*b*) board games. His friend, Patrice, has 6 less than 3 times the number of board games that Terrence has. Write an algebraic expression to represent the number of board games that Patrice has.	8 Jezak carries a backpack to school with a science book that weighs 2 pounds, a math book that weighs 5 pounds, and 4 notebooks each weighing (*p*) pounds. Write an algebraic expression representing the total weight of these books.
9 Willow sold (*v*) video games to the local video store and got \$75.25 in return. She received the same amount for each game. Write an expression to show how much money she earned for each video game.	10 Jayden received \$45 for each lawn (*L*) that he mowed over the summer. Write an expression to represent how much money Jayden earned mowing lawns this summer.

Express It! Numbers, Variables, and Symbols Cards A

Numbers and Variables Cards

32	2	4	*n*	3
10	*n*	2	2	*n*
6	3	3	*n*	*n*
2	*n*	5	*n*	7
2	*n*	8	*n*	3
4	*n*	8	4	*n*
11	2	6		

Symbols Cards (1 set per team)

+	−	*	÷	^
(	)	___		

Express It! Numbers, Variables, and Symbols Cards B

Numbers and Variables Cards

30	**10**	***h***	**400**	**3**
95	***d***	**500**	**20**	***y***
100	**8**	***p***	**55**	**25.95**
5	***t***	**3**	***w***	**6**
2	**5**	**4**	**75.25**	***v***
45	***L***	***w***	***p***	

Symbols Cards (1 set per team)

+	−	*	÷	^
(	)	___		

Express It! Recording Sheet

Name(s): ______________________________ **Date:** ________________

Card Set Used: __________

Card #	Expression

Express It! Directions

Directions

Goal: Given target expressions (in words), create as many expressions (in algebraic form) as possible before all of the cards are used.

- The Written Expressions Cards are shuffled and placed facedown in a deck. The Numbers and Variables Cards are arranged on the playing area faceup in an array. Each team is given a set of Symbols Cards.
- Teams alternate turning over a card from the deck.
- Each team tries to use the number and variable cards as well as its own symbol cards to represent the written expression, without moving the cards. When a team believes it has done so, its players call "Express it!"
- At that point, the team arranges the number and variable cards, and its symbol cards, to make the expression. The other team checks for accuracy.
- If the team is correct, it keeps the number and variable cards it used as a set, retrieves its symbol cards for reuse, and records the expression on its recording sheet. If it is not correct, the team must put the number and variable cards back into the array, retrieve its symbol cards for reuse, and then both teams keep looking for a correct expression.
- After a correct set is made, a team turns over a new Written Expression Card and play continues.
- When there are no more number and variable cards remaining or when no more sets can be made, the team with the greater number of card sets wins the game.

Linear Systems Bingo Game Board

Make your choice of where to place a FREE space.
Then randomly choose the cells in which to write the following answers: (1, 3); (12, 3.5); (2, 4); (2, 9); (3, 2); (6, 4); (9, 22); (2, 3); (5, 7); (1, 1); (12, 3); infinite solutions; infinite solutions; no solution; and no solution.

Linear Systems Bingo Cards

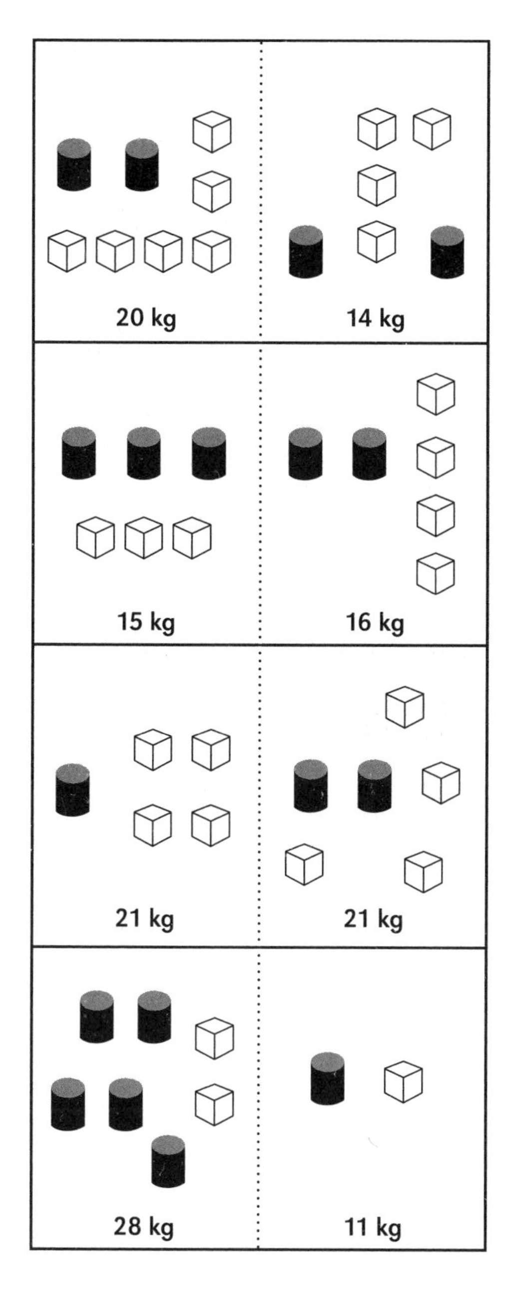

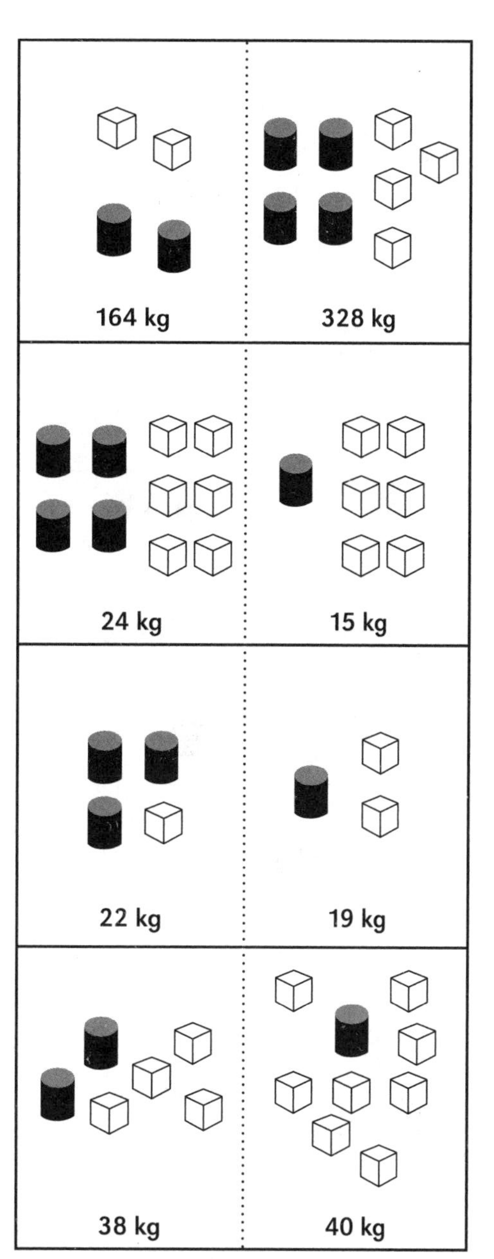

Linear Systems Bingo Cards *(continued)*

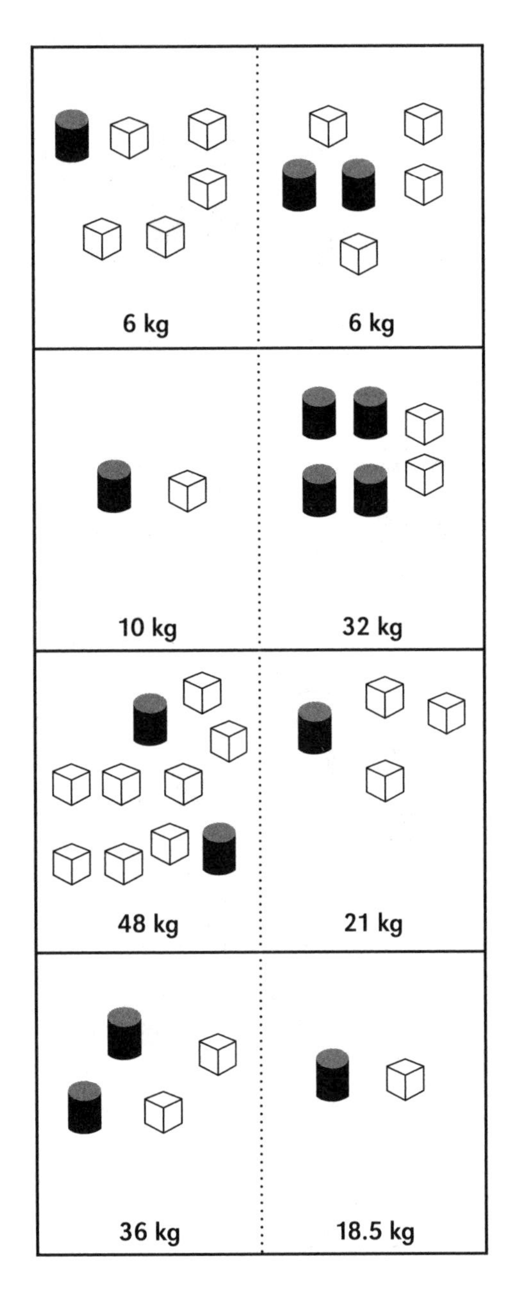

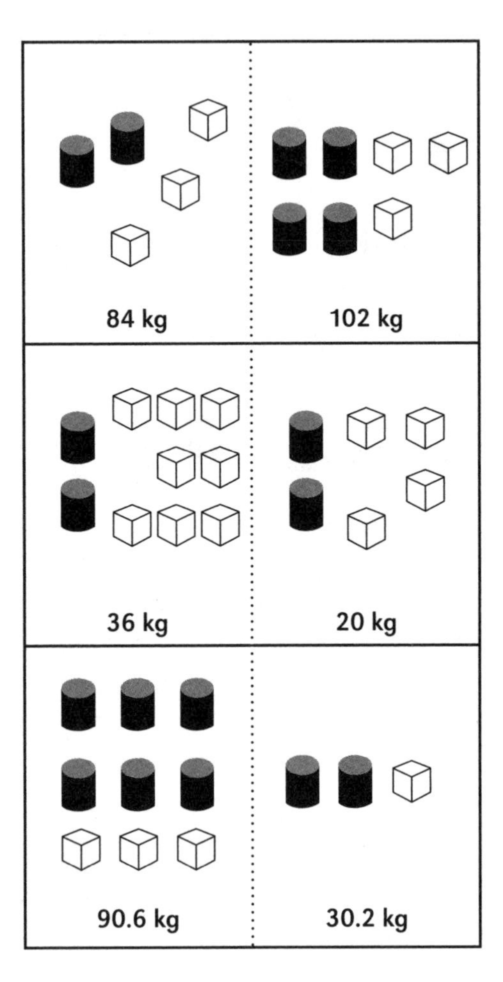

Linear Systems Bingo Directions

Directions

Goal: Be the first to get four answers, or three answers and a FREE space, in a row, column, or diagonal.

- Each team privately completes its copy of the game board by writing *FREE* in one space and then randomly writing the solutions provided in the blank boxes.
- Shuffle the *Linear System Bingo* cards and place them faceup between the teams in a three-by-five array.
- One team chooses a card. Note that the mass of the same shape is equivalent on one card, though not necessarily equivalent to the mass of the same shape on another card.
- Both teams find the solution and discuss the answer and their solution strategies. Note that, though either shape could represent either coordinate, in this game the cylinders will always represent the *x*-coordinate and the cubes will always represent the *y*-coordinate.
- When all players agree, the teams cross out that answer on their game boards and the card is turned facedown.
- Teams alternate choosing a card. The first team to get four answers in a row, column, or diagonal says, "Bingo."
- If both teams agree that all of the answers crossed out are valid, the team that said "Bingo" wins.

Target Statistics Recording Sheet

Name(s): ______________________ **Date:** ______________

Round 1:	Target	Five Data Points	Our Statistic	Points
Mean				
Median				
Range				

Round 2:	Target	Five Data Points	Our Statistic	Points
Mean				
Median				
Range				

Round 3:	Target	Five Data Points	Our Statistic	Points
Mean				
Median				
Range				

Our team's score: ______________

Target Statistics Directions

Directions

Goal: Choose cards to form data sets closest to three "target" statistics: the mean, median, and range.

- Shuffle the cards.
- Turn over the top three cards of the deck for all players to see. The first card identifies the target mean; the second card gives the target median; and the third, the target range. Each team writes these "target" statistics on its recording sheet.
- A player from one team turns over the next card of the deck and each team decides privately whether to record the number in the data set for the mean, median, or range. Once a number is written, it may not be changed.
- Once both teams have recorded the number, the other team turns over a card and again, teams decide where to write this number.
- Cards are turned over and the numbers are recorded until each statistic has exactly five data points.
- Each team determines the mean, median, and range for the data sets it created.
- If one team's statistic is closer to the target, it earns 1 point. If there is a tie, then both teams earn 1 point.
- The team with more points after three rounds wins, or there is a tie.

Data Sense Puzzle A

Name(s): ______________________________ **Date:** ______________

Write the numbers in the blanks so that the math makes sense.

$16\frac{2}{3}$	2
11	3.5
1	30
3	5
50	1
3.5	

Amy was interested in knowing about how many hours a week (to the nearest hour) her classmates play video games on school nights. She surveyed the ______ students in her class (including herself), but she can no longer find her list of the data. She does remember that ______ percent, or ______ students, did not play at all. She knew the range was ______, the median was ______, and the mean was ______. She also found a note she wrote reminding her to tell her mom that she and her best friend were the only ones in her class who play ______ hours per week, or______ hour less than the mean, and that ______ percent of her class play more than ______ hours per week. Later, she remembered that only______ classmate reported playing the greatest number of hours. That response was really an outlier, as no one chose the four hourly choices before it.

Fortunately, Amy finally remembered where she wrote her data!

Data Sense Puzzle B

Name(s): ______________________________ **Date:** ______________

Write the numbers in the blanks so that the math makes sense.

4	**9**
20	**6**
1	**25**
5	**6.5**
6.1	**5**
8	**20**

Liam's coach asked him to poll _____ players at a regional basketball tournament to find out how many hours per week their team practiced. The coach told Liam to get a representative sample by including exactly one player from each of the _____ teams at the tournament. Liam found that the fewest number of hours a team practiced per school day was _____ —or forty-eight minutes— and the greatest was _____, making the range _____. When Liam displayed the data in a box plot, he found the median to be _____. There were _____ responses, or _____ percent of those polled, that were in the lowest quartile range. Liam then calculated the mean to be _____, which was the actual response of _____ player. If instead, that player had given a response of _____ hours, the mean would have been _____.

When Liam shared his report, the coach asked, "So how many hours do you think we should practice?" How many hours do you think Liam should recommend? Why?

Data Sense **Directions**

Directions

Goal: Accurately write the numbers in the blanks in the paragraph.

- Work with a partner.
- Begin by cooperatively reading the paragraph, recognizing the missing numbers as statistical data.
- Interpret the details of the puzzle by using reasoning and sense-making, and calculating the statistics.
- Place the numbers from the box into the blanks. Each number in the box is used exactly once.
- Reread the paragraph to make sure it makes sense.

Frequency Count Puzzle A

What's the Count?

Tamesha observed what 50 people ordered for the Beach Hut Special.

Today's Beach Hut Special
$5.00
Salad Wrap or Pizza Slice
and
Soda, Lemonade, or Water

She found that:

- 32% of the people chose a pizza slice and soda
- 14 more people chose a pizza slice than a salad wrap
- 40% of the people chose soda
- 11 people chose water and a pizza slice
- Two-thirds of the people who chose a salad wrap chose water

How many people chose a salad wrap and lemonade?

Frequency Count Puzzle B

Who Went to What?

Carlos observed people buying 80 tickets to these movies. He surveyed the ticket buyers to see if they were buying tickets for people 12 years or younger, 13–18 years old, or 19 years and older.

Movie Times

4:00 Toy Forest
4:10 Party with Zombies
4:20 Summer of Dance

He found that:

- 27 tickets were bought for *Toy Forest* and none of them were for people 13–18 years old
- 25% of all the tickets sold were for people 12 years or younger
- Three-eighths of all the tickets sold were for *Party with Zombies*
- 35% of all the tickets bought were for people 13–18 years old to see *Party with Zombies*
- Of the tickets sold for *Summer of Dance*, none were for people 12 years or younger and one more ticket was bought for people 19 years and older than for those 13–18 years old
- The ratio of tickets bought for *Toy Forest* for those who were 19 years and older to those who were 12 years or younger was 1 to 2

1. How many tickets were bought for people 19 years and older to see *Party with Zombies*?
2. Do you think the movie ticket bought is related to the age of the person seeing the movie? Explain your thinking.

Frequency Count Directions

Directions

Goal: Answer the question(s), as a result of interpreting the information provided in the puzzle.

- Work alone or with a partner.
- Read the clues.
- Make notes so that you can recall your thinking.
- You do not need to use the clues in order.
- Check your solution with each clue.

I'll Take My Chances Card Set

A Both coins land on heads when tossing two pennies	B A multiple of 2 when picking one card from the deck
C A sum of 6 when rolling two dice	D A 3 or a 4 on either die when rolling two dice
E A black card that is less than 5 when picking one card from the deck	F An odd number on each die when rolling two dice
G At least one penny lands on heads when tossing two pennies	H A red card and a black card when picking two cards from the deck

I'll Take My Chances Card Set *(continued)*

I A number less than 4 when rolling a die or heads when tossing a penny	J A red card when picking one card from the deck and heads when tossing a penny
K A black card when picking one card from the deck and a number greater than 2 when rolling a die	L A multiple of 2 when picking one card from the deck and a number less than 5 when rolling a die
M An even sum when rolling two dice and a face card when choosing one card from the deck	N An odd sum when rolling two dice and heads when tossing a penny
O A face card when choosing one card, heads when tossing a penny, and not a multiple of 2 when rolling a die	P A black card when choosing one card, a number greater than 2 when rolling a die, and heads when tossing a penny

I'll Take My Chances Recording Sheet

Card: ____________ Number of trials it took to get it: ____________

____________ ____________

____________ ____________

____________ ____________

____________ ____________

____________ ____________

____________ ____________

Total number of trials: ____________

Which team got all the "chances" first? ____________

Card: ____________ Number of trials it took to get it: ____________

____________ ____________

____________ ____________

____________ ____________

____________ ____________

____________ ____________

____________ ____________

Total number of trials: ____________

Which team got all the "chances" first? ____________

I'll Take My Chances Directions

Directions

Goal: Get all six outcomes with the fewer number of trials.

- Place all the *I'll Take My Chances* cards facedown between the two teams.
- Choose six cards to use for the game and mark the letter of each card chosen in the same order on each team's recording sheet.
- Take turns.
- On its first turn, Team 1 reads the situation written on the card and both teams agree on the expectations. Team 1 then carries out the simulation.
- If the team is able to match the event in one try, it checks it off on the recording sheet. On its next turn, Team 1 will try to match the event on the next card.
- If the team cannot match the event, its turn ends and the team makes a tally mark on its recording sheet to indicate that it tried the event. Team 1 must try again to match the event on its next turn.
- The teams then alternate turns, trying to match the outcomes described in the order they appear on the recording sheet.
- The first team to match all outcomes wins the game.

Is It a Match? Card Set

Picking a number other than 4 from cards 1–10 placed facedown	$\frac{9}{10}$	**Tossing a coin and getting tails**
Landing on D when dropping a paper clip in the rectangle N | HD P | Q | D	$\frac{1}{8}$	$\frac{1}{2}$
Picking a blue tile from a bag Color: brown 1; green 1; blue 2; pink 5	$\frac{2}{9}$	**Picking, with eyes closed, a shaded block**
Spinning a 2 Spinner sections: 2, 2, 2, 3, 4, 4, 1, 1	$\frac{3}{8}$	$\frac{4}{6}$

Is It a Match? Card Set *(continued)*

<table>
<tr>
<td>Tossing a coin three times and getting heads each time</td>
<td>$\frac{1}{8}$</td>
<td>Have the day of the week start with the letter "S" or "T"</td>
</tr>
<tr>
<td>Picking a red card or an ace from a deck of playing cards</td>
<td>$\frac{28}{52}$</td>
<td>$\frac{4}{7}$</td>
</tr>
<tr>
<td>Picking a number less than 16 from cards 11–20 placed facedown</td>
<td>$\frac{1}{2}$</td>
<td>Picking, with eyes closed, a shaded rectangle or a circle</td>
</tr>
<tr>
<td>Picking a tile that isn't green from a bag
<table>
<tr><th>Color</th><th>Number</th></tr>
<tr><td>brown</td><td>1</td></tr>
<tr><td>green</td><td>1</td></tr>
<tr><td>blue</td><td>2</td></tr>
<tr><td>pink</td><td>5</td></tr>
</table>
</td>
<td>$\frac{8}{9}$</td>
<td>$\frac{4}{7}$</td>
</tr>
</table>

Is It a Match? Card Set *(continued)*

Rolling a 4 on a die, and getting heads when tossing a coin	$\frac{1}{12}$	**Picking a number that isn't odd from cards 11–20 placed facedown**
Spinning two spinners and landing on 2 on the first and 1 on the second	$\frac{1}{4}$	$\frac{1}{2}$
Rolling a red die and getting an odd number, and rolling a blue die and getting a multiple of 2	$\frac{1}{4}$	**Rolling a die and getting a 4, and spinning the spinner and landing on pink**
Rolling a die and getting a 4, and tossing two coins and getting two heads	$\frac{1}{24}$	**0**

Is It a Match? Directions

Directions

Goal: Get the most card sets when all cards have been matched.

- Shuffle all the cards and place them facedown in a 4-by-6 array between the two teams.
- On each turn, the team:
  1. Turns over two cards.
  2. The two cards match if one card represents an event and the other card gives the probability of its occurrence. If the cards match, the team keeps the cards and turns over another two cards.
  3. If the cards do not match, the team must return the cards facedown in the same place they were in the array.
- The game is over when all the cards have been matched.
- The winner is the team who has the most matched card sets.

Patternary Card Set

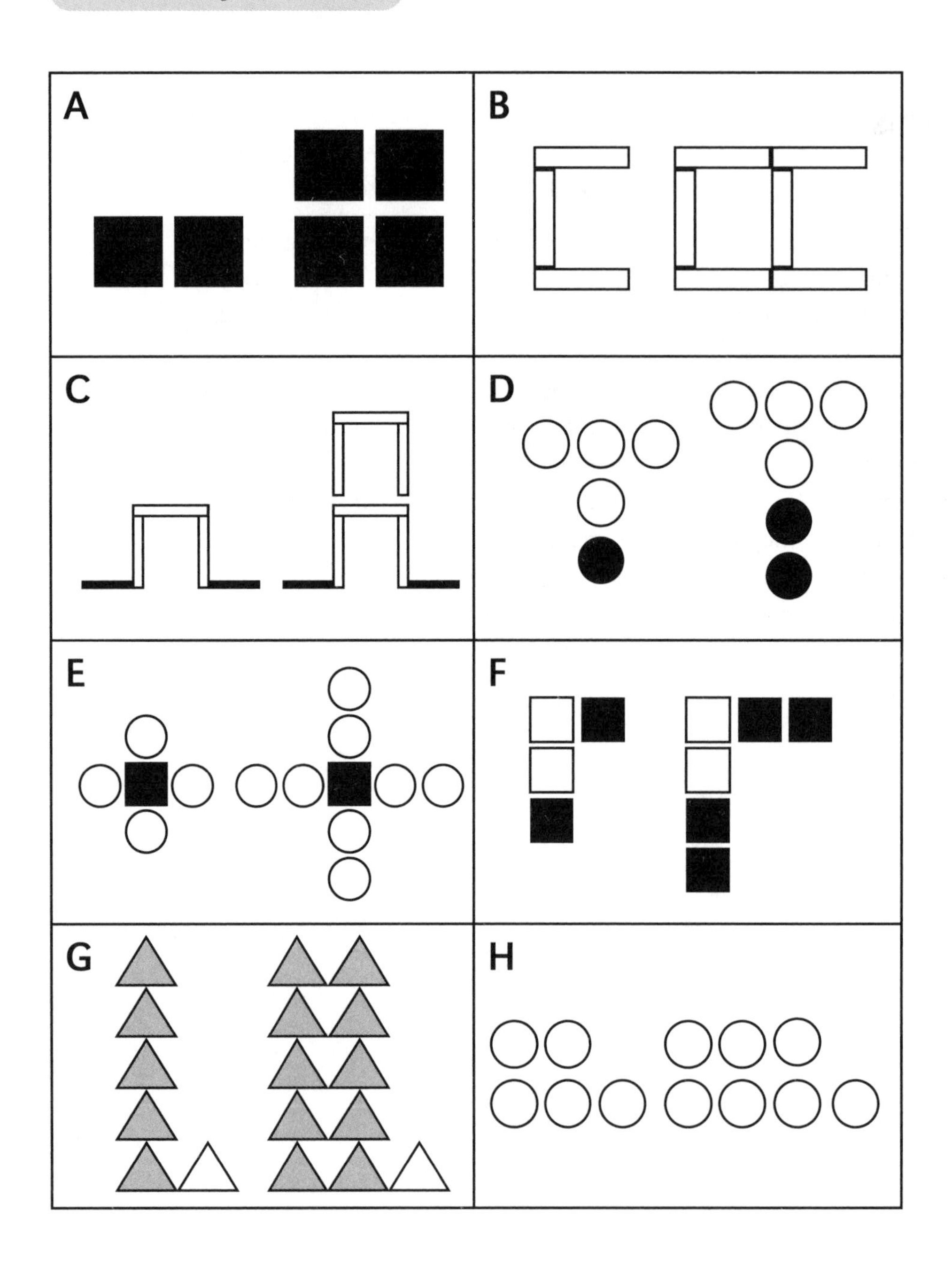

Patternary Card Set *(continued)*

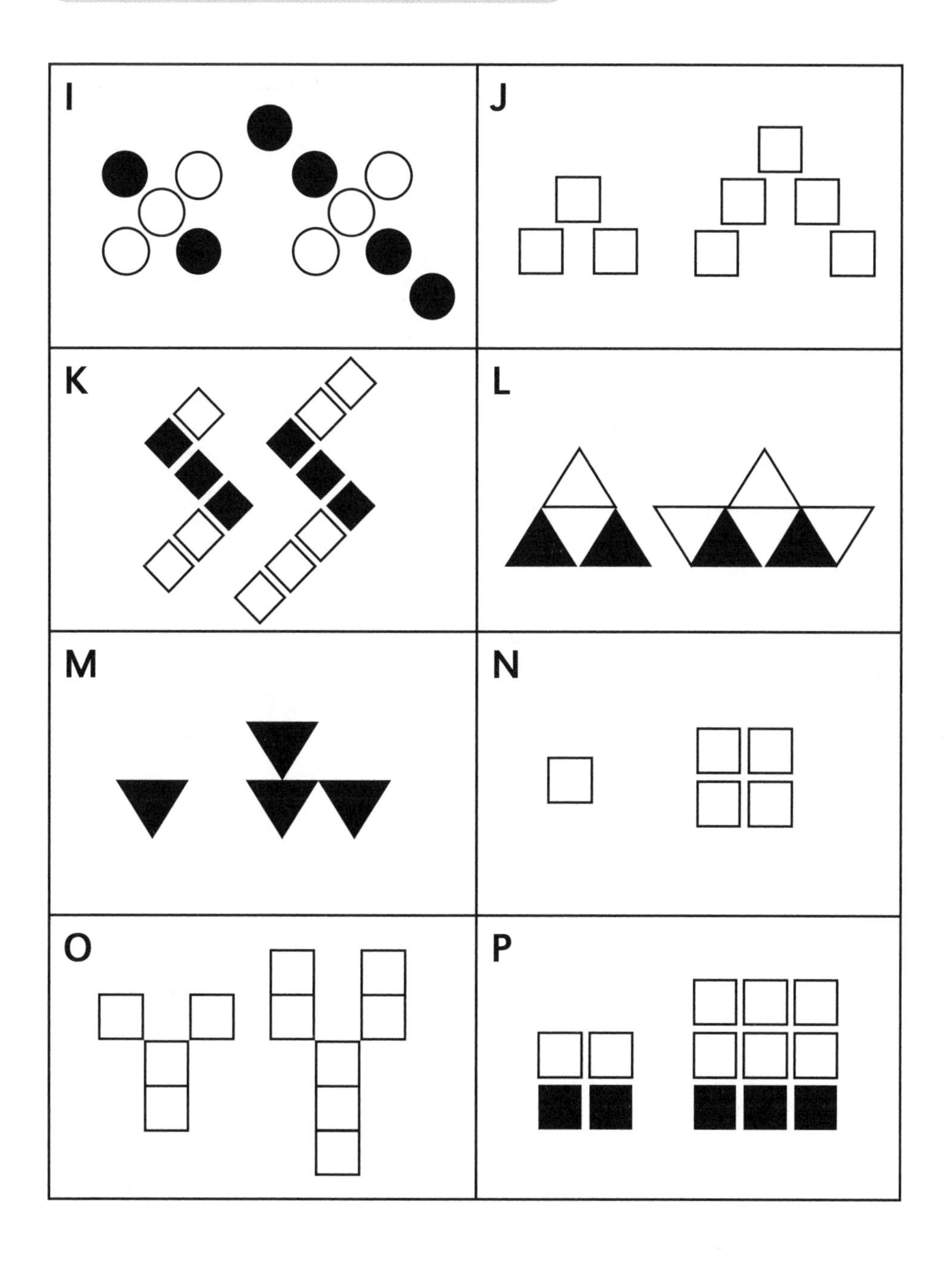

Patternary Recording Sheet

Name(s): ______________________ **Date:** __________

Draw the next four images.
Fold the page so that you show only one new image each time.

Round 1 **Card:**	**Round 2** **Card:**

Equation or description:

Equation or description:

Patternary Directions

Directions

Goal: Play cooperatively to earn a total of 40 points by identifying and continuing growing patterns.

- The *Patternary* Card Set is shuffled and placed facedown between the two teams.
- For each round of the game, each team:
  1. Chooses a card from the card set, considers the images on the card, and thinks about how the pattern might be growing;
  2. Identifies a growing pattern and continues it by drawing the next four images on the recording sheet;
  3. Writes an equation or a description to represent the growing pattern.
- Team 1 shows Team 2 players the *Patternary* card it chose. After seeing the card, Team 2 gets one try to provide a description or an equation (or an equivalent description or equation) that matches the growing pattern Team 1 identified. If Team 2 cannot do so, players ask to see the first image Team 1 drew to extend the pattern.
- Play continues with Team 1 showing one drawing at a time and Team 2 being given one try to identify the pattern each time a new drawing is shown.
- If Team 2 is able to provide a description or identify the equation by the time the fourth student-drawn image is shown, both teams earn 15 points, minus 1 point for each drawing shown. If Team 2 is unable to identify the equation after seeing the original card and the four drawings, it is asked to make a drawing that fits the pattern. If Team 1 agrees that the drawing is correct, both teams earn 5 points.
- If one team has recorded a description and the other an equation, teams talk until they agree there is a match.
- Teams reverse roles and complete the round.
- After two rounds, both teams are considered to be winners if the group has earned 40 points.

What's My Rule? Card Set

$x + 4 = y$	$x + 1 = y$	$2x - 1 = y$
$3x - 1 = y$	$2x + 5 = y$	$2x + \frac{1}{2} = y$
$x^2 = y$	$\frac{1}{2}x - 1 = y$	$2x^2 = y$
$2(x - 1) = y$	$\frac{(x + 4)}{2} = y$	$x^2 + 2 = y$

What's My Rule? Recording Sheet

Name(s): ______________________ **Date:** ____________

Turn 1:

Input (x)	**Output** (y)

Turn 2:

Input (x)	**Output** (y)

Turn 3:

Input (x)	**Output** (y)

The rule is: ______________

The rule is: ______________

The rule is: ______________

What's My Rule? Directions

Directions

Goal: Identify rules by using fewer input values than the opposing team uses to identify its rules.

- The *What's My Rule?* cards are shuffled and placed facedown as a deck between the two teams.
- Team 1 chooses a card from the deck and keeps it private, as the card identifies the rule for the turn.
- Team 2 gives Team 1 a number to input (x) into the rule and then Team 1 tells Team 2 the output number (y).
- Team 2 tries to guess the rule, in the form of an equation.
- Team 2's score is the number of input values it names before identifying the rule.
- It is then Team 2's turn to choose a card from the deck.
- At the end of three turns for each team, the team with fewer points is the winner.

Graph Story Puzzle A

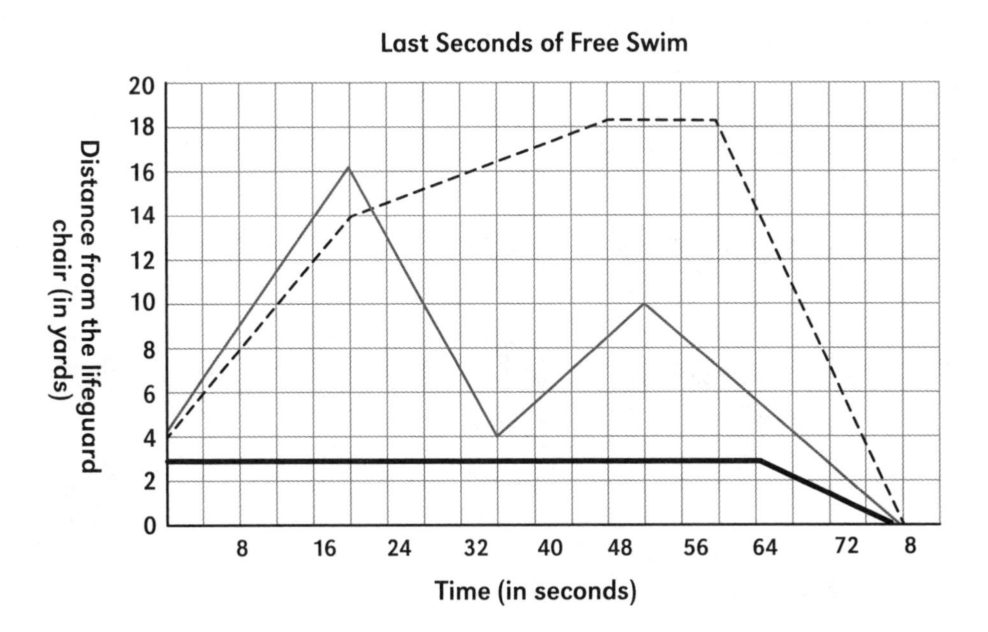

Zak, Grace, and Horatio were at City Camp. Their favorite activity was free swim. During the last 80 seconds of today's free swim, there were times when Grace was farther from the lifeguard chair than Horatio was. When there were 40 seconds left of free swim, Horatio was farther from the lifeguard chair than Zak was. Horatio stopped swimming for 12 seconds to fix his goggles and then swam at his fastest rate. The campers had all returned to the lifeguard chair when free swim ended.

Use the graph and the story to help you decide how fast Grace was swimming between 20 and 36 seconds. ___________

Graph Story Puzzle B

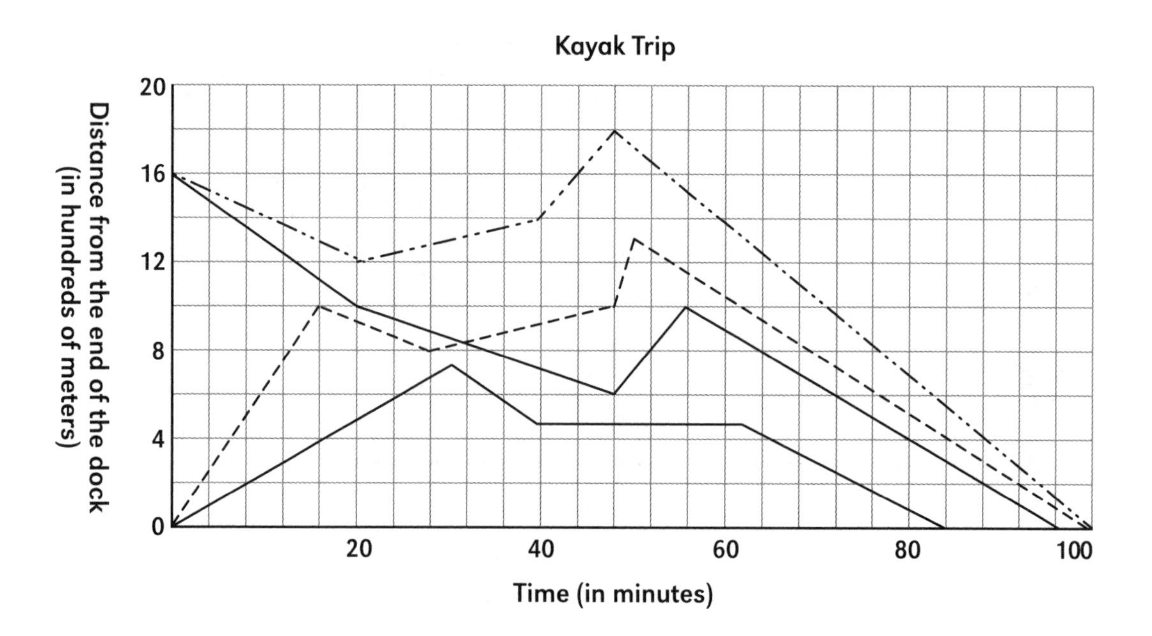

Mark, Phillippe, Holly, and Nila chose kayaking for their first activity at City Camp. Each kayaker carried a device with an app that let the counselors keep track of how far the campers were from the end of the dock. Two of the campers started out at the dock and the other two started at the tiny island that was in the lagoon. Three of the campers practically crashed into each other. Holly paddled away and then stopped for several minutes to calm down after the near crash. Then she headed for the dock. For a while, Mark and Nila were paddling at the same speed. When Phillippe realized he was headed toward Mark's kayak, he turned and headed for the dock. They all returned to the dock so that they could look at their trips on the app.

Use the graph and the story to help you find Nila's average speed during these 100 minutes. ___________

Graph Story Directions

Directions

Goal: Match each part of the graph with one camper's story and answer the related question.

- Read the story as it relates to the graph or look at the graph and relate it to the story.
- Determine which part of the graph matches which camper's story.
- Answer the question.
- Be sure to check your work.

Function Connection Card Set

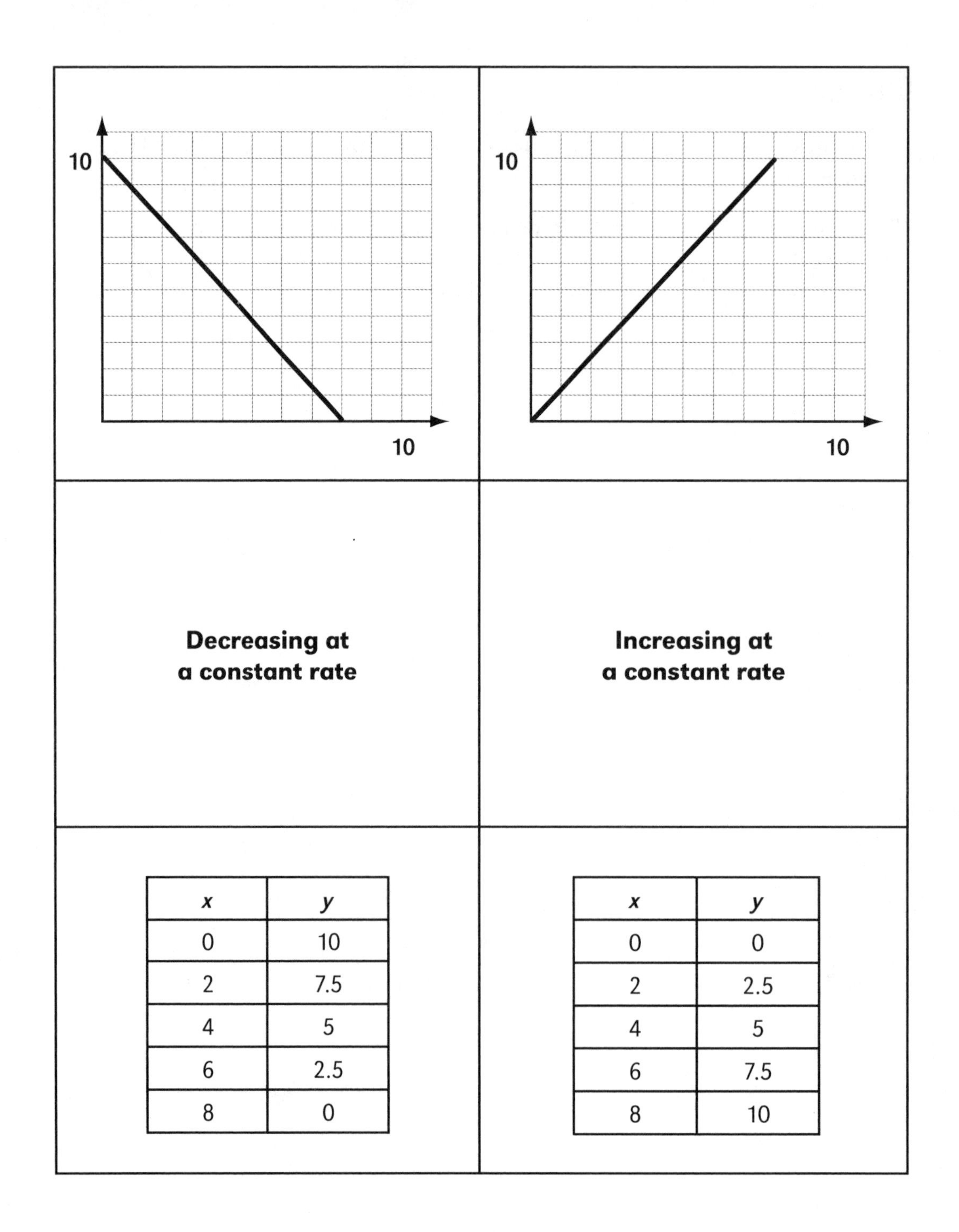

Decreasing at a constant rate

Increasing at a constant rate

x	*y*
0	10
2	7.5
4	5
6	2.5
8	0

x	*y*
0	0
2	2.5
4	5
6	7.5
8	10

Function Connection Card Set *(continued)*

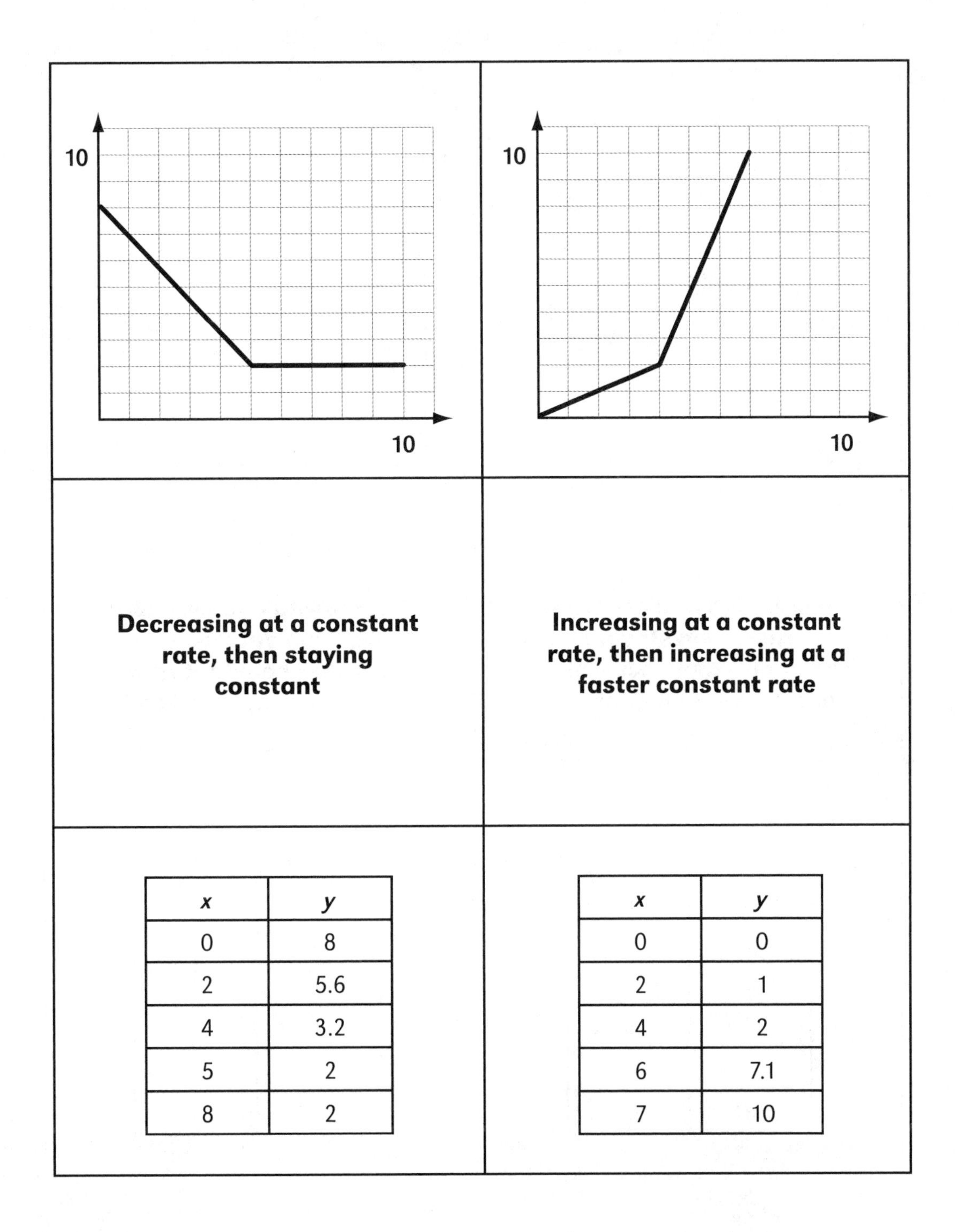

x	y
0	8
2	5.6
4	3.2
5	2
8	2

x	y
0	0
2	1
4	2
6	7.1
7	10

Function Connection Card Set *(continued)*

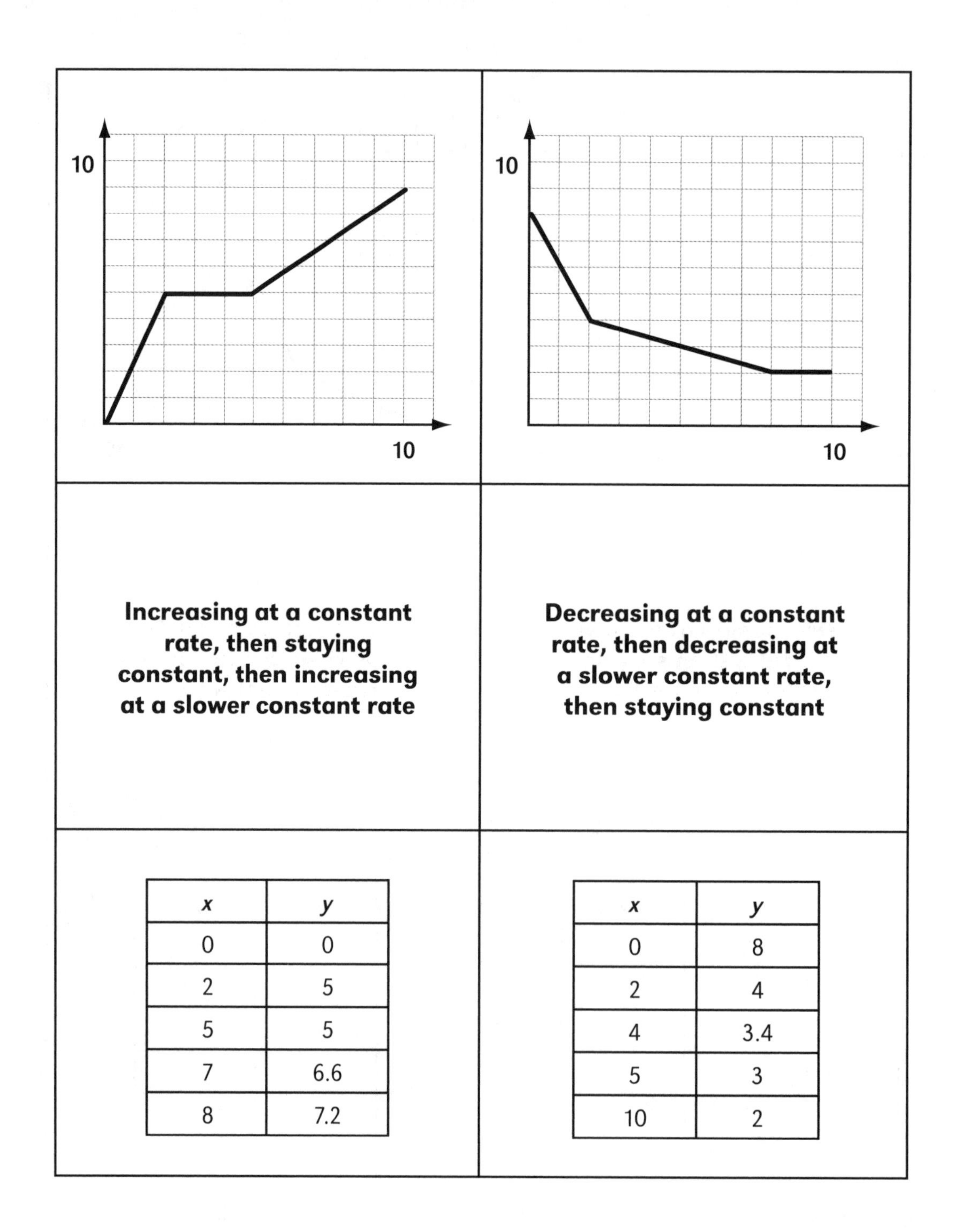

x	y
0	0
2	5
5	5
7	6.6
8	7.2

x	y
0	8
2	4
4	3.4
5	3
10	2

Function Connection Card Set *(continued)*

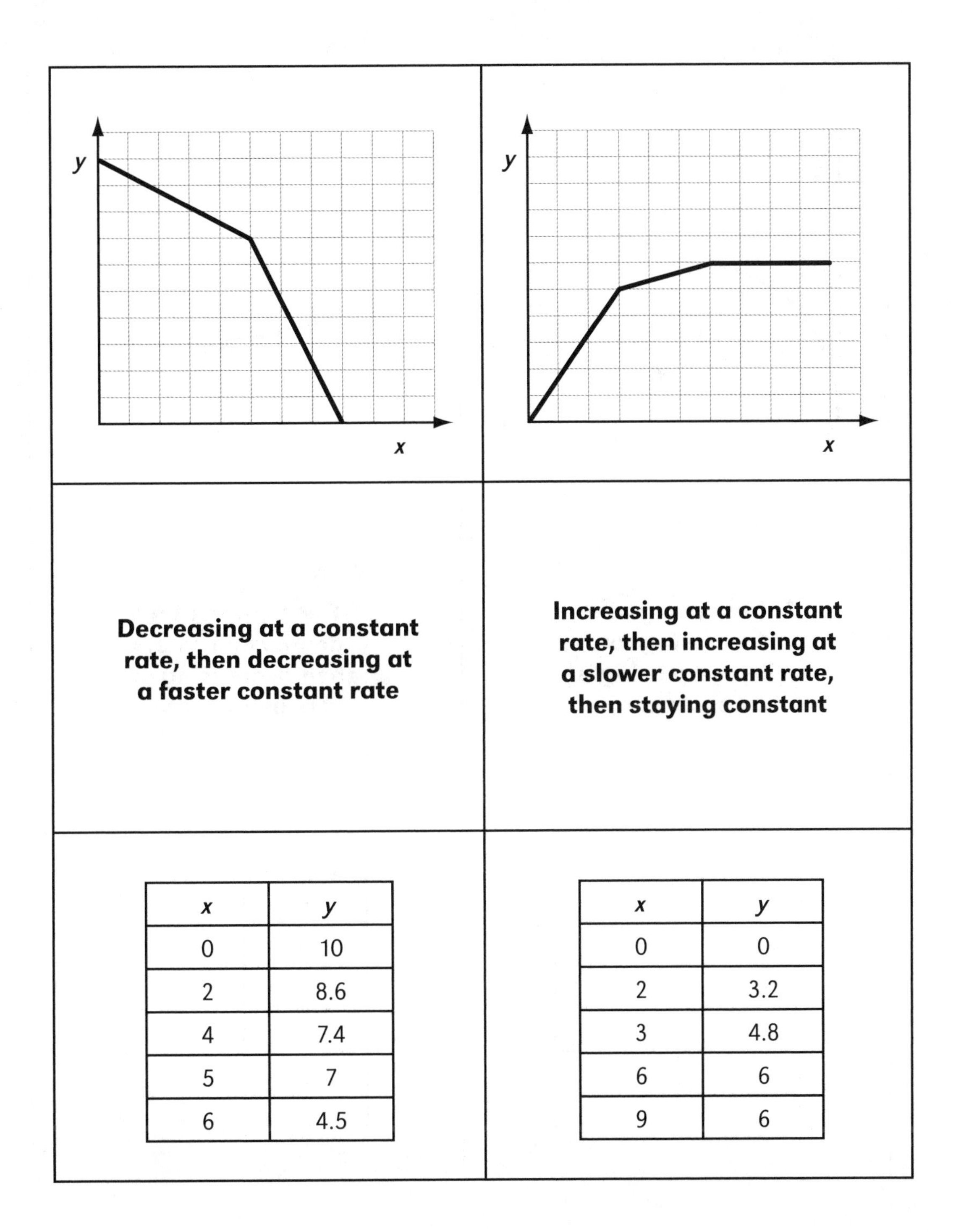

x	y
0	10
2	8.6
4	7.4
5	7
6	4.5

x	y
0	0
2	3.2
3	4.8
6	6
9	6

Function Connection Card Set *(continued)*

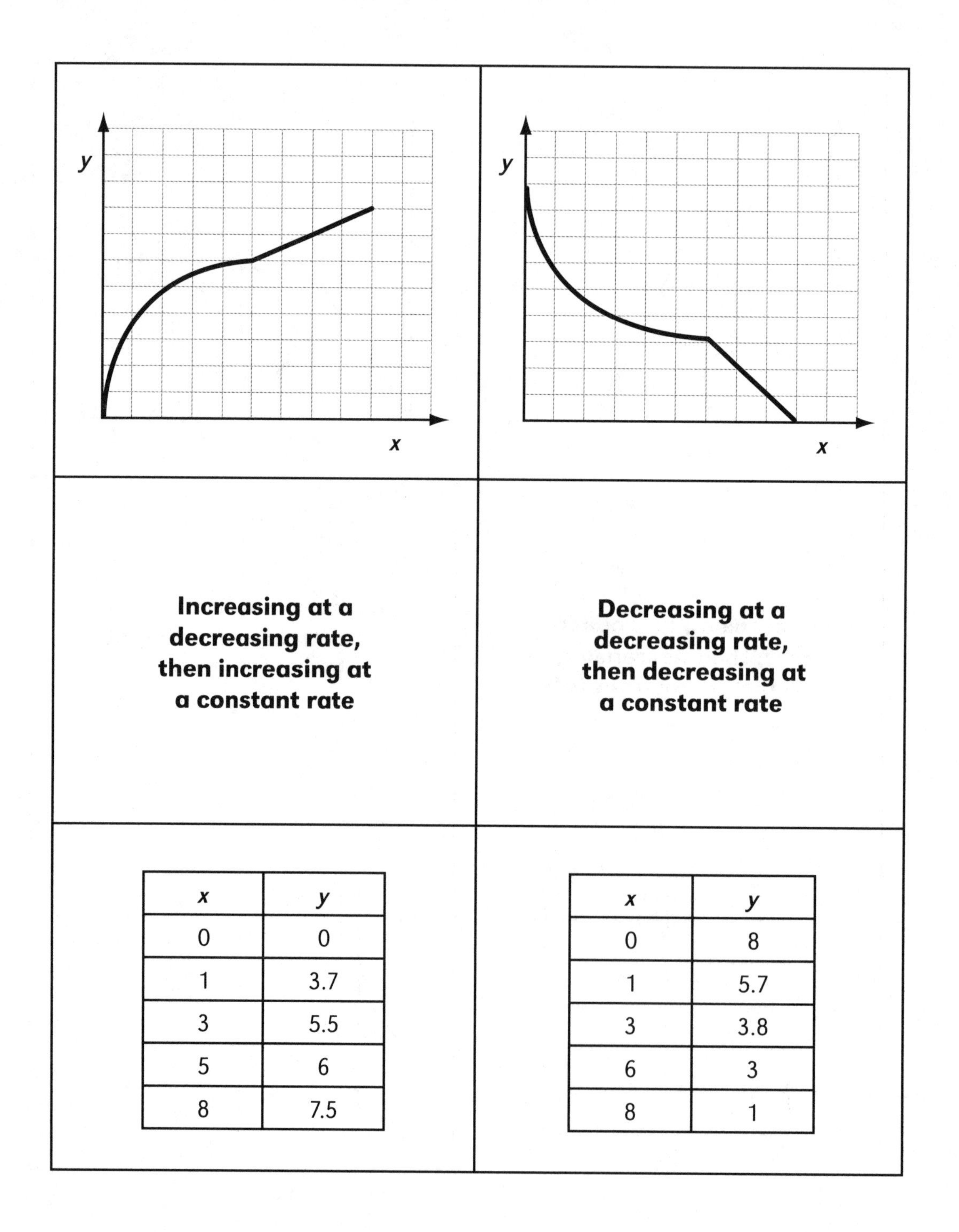

x	y
0	0
1	3.7
3	5.5
5	6
8	7.5

x	y
0	8
1	5.7
3	3.8
6	3
8	1

Function Connection Directions

Directions

Goal: Get the most points by matching graphs, tables, and descriptions.

- Identify one player to also be the scorekeeper.
- Deal four cards to each team. Place the other cards facedown in a deck.
- Whenever a team has two or three cards in its hand that match and the other team agrees, the players put the matching set between both teams for all to see. The scorekeeper records one point for each card in the set for that team.
- Whenever a team has one card in its hand that matches a set with two cards that have already been played and the other team agrees, it adds that card to the set and the scorekeeper records one point for that team.
- Take turns being Team 1 and Team 2.
- On each turn, Team 1 begins by asking Team 2 to choose a card from its hand to show faceup. Then:
  1. If Team 1 has one or two cards in its hand that match that card, it shows the card(s) to Team 2. If Team 2 agrees, Team 1 players make a set placing all the cards together faceup, for everyone to see. The scorekeeper records one point for each card in the set.
  2. If Team 1 believes the card matches a set already made, belonging to either team, it shows the match to Team 2. If Team 2 agrees, Team 1 places the card with that set and the scorekeeper records one point.
  3. If Team 1 cannot make a match with the card, Team 2 takes the card back and Team 1 draws a card from the deck.
- When there are no cards left in the deck, play continues without the players drawing cards.
- When teams believe they have made all possible matches, the game ends, points are totaled, and the team with more points is the winner.

Find One Puzzle A

Match each equation with one of its facts. You must use each fact exactly once. Write the letter of the fact under the equation.

Fact List

A. The slope is 5.	B. Non-linear	C. The value of x is 6 when y is 0.	D. The value of y is -3 when x is 0.
E. The point (5, 25) is a solution.	F. The value of y is -16 when x is 0.	G. The slope is -2.	H. The point (-2, 0) is a solution.
I. The value of x is -4 when y is 0.	J. Linear	K. The point (6, 37) is a solution.	L. The slope is -2 and the point (5, 0) is a solution.

Equation Board

$2x - 10 = -y$ ______	$y - 2 = x$ ______	$y = x^2 + 1$ ______
$-2x - 16 = y$ ______	$y = 5x$ ______	$3(x - 1) = y$ ______
$y + 6 = x$ ______	$x^2 = y$ ______	$5x + 20 = y$ ______
$5x - 3 = y$ ______	$2y + 6 = -4x$ ______	$y = -24x - 15$ ______

Find One Puzzle B

Match each linear equation with one of its facts. You must use each fact exactly once. Write the letter of the fact under the equation.

Fact List

A. The point (2, 6) is a solution.	B. The value of x is 3 when y is 0.	C. The slope is 3.	D. The y-intercept is 0.
E. The slope is negative.	F. The points (8, -33) and (2, -3) are solutions.	G. The y-intercept is 1.	H. The value of x is $-\frac{4}{3}$ when y is 0.
I. The y-intercept is 12.	J. The slope is $\frac{3}{2}$.	K. The value of y is $\frac{7}{2}$ when x is 0.	L. The point (7, 34) is a solution.

Equation Board

$3x + 1 = y$ ______	$3x - 2y = 0$ ______	$y = \frac{3}{4}x + 1$ ______
$x = \frac{-y}{15} + 2$ ______	$y = 3x$ ______	$-(5x - y) = -1$ ______
$\frac{-y}{4} + 3 = x$ ______	$2y + 3x = 0$ ______	$-3x + 12 = y$ ______
$-5x + 7 = y$ ______	$x = \frac{(2y - 7)}{3}$ ______	$y = \frac{4}{3}x + 12$ ______

Find One Directions

Directions

Goal: Match each equation with exactly one fact, using each fact once.

- Each fact in the list matches one or more of the equations on the equation board.
- Write each fact on a blank beneath one of its matching equations on the board. Each fact must be written exactly once.
- Write the facts so that each equation gets a match.
- Be sure to check your work.

Answer Key

Chapter 3

Match It

Each side is matched with its equivalent as displayed on the reproducibles.

Tic-Tic-Tac-Toe

Game Board A:

$315 \div 9 = 35$
$315 \div 35 = 9$
$315 \div 105 = 3$
$315 \div 15 = 21$
$315 \div 3 = 105$

$7{,}875 \div 9 = 875$
$7{,}875 \div 35 = 225$
$7{,}875 \div 105 = 75$
$7{,}875 \div 15 = 525$
$7{,}875 \div 3 = 2{,}625$

$18{,}900 \div 9 = 2{,}100$
$18{,}900 \div 35 = 540$
$18{,}900 \div 105 = 180$
$18{,}900 \div 15 = 1{,}260$
$18{,}900 \div 3 = 6{,}300$

$2{,}835 \div 9 = 315$
$2{,}835 \div 35 = 81$
$2{,}835 \div 105 = 27$
$2{,}835 \div 15 = 189$
$2{,}835 \div 3 = 945$

$1{,}260 \div 9 = 140$
$1{,}260 \div 35 = 36$
$1{,}260 \div 105 = 12$
$1{,}260 \div 15 = 84$
$1{,}260 \div 3 = 420$

Game Board B:

$1\frac{3}{4} \div \frac{3}{8} = 4\frac{2}{3}$
$1\frac{3}{4} \div 3 = \frac{7}{12}$
$1\frac{3}{4} \div 1\frac{1}{4} = 1\frac{2}{5}$
$1\frac{3}{4} \div 4 = \frac{7}{16}$
$1\frac{3}{4} \div \frac{5}{6} = 2\frac{1}{10}$

$6 \div \frac{3}{8} = 16$
$6 \div 3 = 2$
$6 \div 1\frac{1}{4} = 4\frac{4}{5}$
$6 \div 4 = 1\frac{1}{2}$
$6 \div \frac{5}{6} = 7\frac{1}{5}$

$3\frac{1}{2} \div \frac{3}{8} = 9\frac{1}{3}$
$3\frac{1}{2} \div 3 = 1\frac{1}{6}$
$3\frac{1}{2} \div 1\frac{1}{4} = 2\frac{4}{5}$
$3\frac{1}{2} \div 4 = \frac{7}{8}$
$3\frac{1}{2} \div \frac{5}{6} = 4\frac{1}{5}$

$2 \div \frac{3}{8} = 5\frac{1}{3}$
$2 \div 3 = \frac{2}{3}$
$2 \div 1\frac{1}{4} = 1\frac{3}{5}$
$2 \div 4 = \frac{1}{2}$
$2 \div \frac{5}{6} = 2\frac{2}{5}$

$3\frac{1}{3} \div \frac{3}{8} = 8\frac{8}{9}$
$3\frac{1}{3} \div 3 = 1\frac{1}{9}$
$3\frac{1}{3} \div 1\frac{1}{4} = 2\frac{2}{3}$
$3\frac{1}{3} \div 4 = \frac{5}{6}$
$3\frac{1}{3} \div \frac{5}{6} = 4$

Game Board C:

$360 \div 3 = 120$
$360 \div 0.1 = 3600$
$360 \div 0.8 = 450$
$360 \div 0.4 = 900$
$360 \div 6 = 60$

$0.2 \div 3 = 0.0\overline{6}$
$0.2 \div 0.1 = 2$
$0.2 \div 0.8 = 0.25$
$0.2 \div 0.4 = 0.5$
$0.2 \div 6 = 0.0\overline{3}$

$6.3 \div 3 = 2.1$
$6.3 \div 0.1 = 63$
$6.3 \div 0.8 = 7.875$
$6.3 \div 0.4 = 15.75$
$6.3 \div 6 = 1.05$

$40 \div 3 = 13.\overline{3}$
$40 \div 0.1 = 400$
$40 \div 0.8 = 50$
$40 \div 0.4 = 100$
$40 \div 6 = 6.\overline{6}$

$824 \div 3 = 274.\overline{6}$
$824 \div 0.1 = 8{,}240$
$824 \div 0.8 = 1{,}030$
$824 \div 0.4 = 2{,}060$
$824 \div 6 = 137.\overline{3}$

Three-For

Answers will vary based on the target number and what number students choose.

Communicate It!

Not applicable

A-Maz-ing!

Answers will vary based on how students move through the game board.

Chapter 4

Four of a Kind

The first and second rows of cards, the third and fourth rows of cards, the fifth and sixth rows of cards, and so forth, are sets of four matching cards.

Best Unit Price

The cards for each product are presented in order from least expensive (best) to most expensive unit price.

Keep Going?

Introductory cards:
Card A: Run 15 minutes at 6 mph, distance = $1\frac{1}{2}$ miles
Card B: Swim 20 minutes at 1 mph, distance = $\frac{1}{3}$ mile
Card C: Bike 5 minutes at 12 mph, distance = 1 mile
Card D: Jog 10 minutes at 5 mph, distance = $\frac{5}{6}$ mile
Card E: Walk 15 minutes at 3 mph, distance = $\frac{3}{4}$ mile
Card F: Rollerblade 10 minutes at 10 mph, distance = $1\frac{2}{3}$ miles

Card set:
Skateboard 40 minutes at 7.5 mph, distance = 5 miles
Rollerblade 20 minutes at 10 mph, distance = $3\frac{1}{3}$ miles
Walk 40 minutes at 3.5 mph, distance = $2\frac{1}{3}$ miles
Run 30 minutes at 6 mph, distance = 3 miles
Jog 75 minutes at 4 mph, distance = 5 miles
Bike 20 minutes at 12 mph, distance = 4 miles
Unicycle 15 minutes at 2 mph, distance = $\frac{1}{2}$ mile

Car 5 minutes at 60 mph, distance = 5 miles
Skateboard 30 minutes at 12 mph, distance = 6 miles
Rollerblade 40 minutes at 10 mph, distance = $6\frac{2}{3}$ miles
Walk 60 minutes at 4 mph, distance = 4 miles
Run 30 minutes at 5 mph, distance = $2\frac{1}{2}$ miles
Jog 45 minutes at 4 mph, distance = 3 miles
Bike 24 minutes at 15 mph, distance = 6 miles
Unicycle 30 minutes at 3 mph, distance = $1\frac{1}{2}$ miles
Car 10 minutes at 42 mph, distance = 7 miles

The Question Is/The Answer Is

Introductory puzzle:
A: $240, B: $200, C: $245, D: $240; order as A, B, D, C
The puzzle pieces are presented in order, with the answer to the question on the last card found on the first card.

Money Matters

Introductory puzzle:
The $350 TV had a 35% discount, the $425 TV had a 25% discount, and the $500 TV had a 20% discount.

Puzzle A:
Lewis had $280, Marie had $40, Norah had $200, Odette had $72, and Portia had $200.

Puzzle B:
Andy $660; Bwan $500; Carry $925; Dansi $660; Ethan $595.50

Chapter 5

Find It Together

Mini-puzzle: Bertram was 90 minutes late. $(\frac{27\text{-}18}{3})30 = x$
Puzzle A: Mayara had 6 pencils. $\frac{x + 6 + 3x}{2} + 5 = 20$
Puzzle B: Giovanni's allowance is $18. $2x - \frac{1}{2}x - 5 + 7 = 29$

Equivalent Expressions

page A-41
$-(x - 12)$ and $12 - x$
$2x - 2 + 5$ and $3 + 2x$
$2x - (-2) + 24$ and $2(x + 13)$

$7x + 5x - 2$ and $-2 + x(7 + 5)$
$2x - 7$ and $-3 - 4 - (-2x)$
$-3(x - 7)$, $30 - 3(3 + x)$, $20 - 3x - 1 + 2$, and $-3x + 21$
$12x + 15y$, $3(4x + 5y)$, $15y + 12x$, and $-3(-4x + (-5y))$

page A-42

$24x + 72y$ and $(2x + 6y)12$

$9x + 3 - 2x - 15$ and $7x - 12$

$-(7x + 12)$ and $-12 - 7x$

$48 - 3x + 4(x - 9)$ and $x + 12$

$\frac{(35x + 45)}{5}$ and $7x + 9$

$\frac{(24x - 64)}{4}$, $2(3x - 8)$, $-12x - 2(-9x + 4)$, and $-8 + 6x$

$-16x + 12$, $12 - 16x$, $4(3 - 4x)$ and $-(22x + 8 - 6x + (-20))$

page A-43

$\frac{3}{4}x + 1\frac{1}{2}$ and $\frac{3}{4}(x + 2)$

$\frac{-(3 + 6x)}{0.5}$ and $-6(2x + 1)$

$0.4(-10x + 5)$ and $2 - 4x$

$19 - 0.2(10x + 10)$ and $17 - 2x$

$9 - 0.34x + 2x - 0.66x$ and $x + 9$

$-0.75x + 1 + 0.5x - 0.5$, $-0.25x + 0.5$, $-0.5(0.5x - 1)$ and $0.75x + 0.5 + (-x)$

$\frac{-(\frac{1}{4} + \frac{3}{4}x)}{\frac{1}{4}}$, $-(1 + 3x)$, $\frac{-(\frac{3}{2}x + \frac{1}{2})}{\frac{1}{2}}$, and $7.5x - 6 - 10.5x + 5$

Solve It

Answers will vary.

Express It!

Card Set A:

1. $\frac{32}{(2n + 4)}$
2. $3n - 10$
3. $n^2 + 6$
4. $2n - 6$
5. $3(3 + n)$
6. $2n + 5n$
7. $\frac{7^2}{n}$
8. $(8n + 2)^3$
9. $4n - 11$
10. $\frac{8}{(4n)}$

Card Set B:

1. $30h - 10$
2. $400 - 3w$
3. $500 + 95d$
4. $20 + y$
5. $8\frac{p}{100}$
6. $55.25 - (25.95 + 5t)$
7. $3b - 6$
8. $2 + 5 + 4p$
9. $\frac{75.25}{v}$
10. $45L$

Linear Systems Bingo

(aligned to presentation of cards)
(1, 3); infinite solutions
(2, 3); (3, 2)
no solution; (5, 7)
(2, 9); (12, 3.5)
(1, 1); (9, 22)
(6, 4); (2, 4)
(12, 3); infinite solutions
no solution

Chapter 6

Target Statistics

Not applicable

Data Sense

Mini-puzzle: 25, 7.5, 9.5, 2
Puzzle A: 30, $16\frac{2}{3}$, 5, 11, 3.5, 3, 2, 1, 50, 3.5, 1
Puzzle B: 20, 20, 4, 9, 5, 6.5, 5, 25, 6, 1, 8, 6.1

Frequency Count

Puzzle A:

	Salad Wrap	Pizza Slice	Total
Soda	4	16	20
Lemonade	2	5	7
Water	12	11	23
Total	18	32	50

Two people chose a salad wrap and lemonade.

Puzzle B:

	Toy Forest	Party with Zombies	Summer of Dance
≤12	18	2	0
13–18	0	28	11
≥19	9	0	12

1. Zero tickets to *Party with Zombies* were bought for people over 19.
2. Answers will vary, but should indicate examples of a relationship between which tickets were bought and the people's age.

I'll Take My Chances

Not applicable

Is It a Match?

Cards in column 1 match cards in column 2. Cards in column 3 match in sets of two, with the probability given under each outcome.

Chapter 7

Patternary

Note that a variety of patterns are possible, depending on the variables the students choose. Possible answers are provided, all of which relate the total number of shapes in the figure (t) to the figure number (f).

A $t = 2f$
B $t = 3f$
C $t = 3f + 2$
D $t = f + 4$
E $t = 4f + 1$
F $t = 2f + 2$
G $t = 5f + 1$
H $t = 2f + 3$
I $t = 2f + 3$
J $t = 2f + 1$
K $t = 2f + 4$
L $t = 2f + 2$
M $t = 2f - 1$
N $t = f^2$
O $t = 3f + 1$
P $t = (f + 1)(f + 1)$

What's My Rule?

Answers are the equations on the cards or the equations the teams create.

Graph Story

Puzzle A: $\frac{3}{4}$ yard per second

Puzzle B: 28 meters per minute

Function Connection

The three cards in each column match.

Figure 7.7 Do Now Task

Answers: F, T, F, F, F

Find One

Puzzle A:

$2x - 10 = -y$ is L

$-2x - 16 = y$ is F

$y + 6 = x$ is C

$5x - 3 = y$ is A

$y - 2 = x$ is H

$y = 5x$ is E

$x^2 = y$ is B

$2y + 6 = -4x$ is G

$y = x^2 + 1$ is K

$3(x - 1) = y$ is D

$5x + 20 = y$ is I

$y = -24x - 15$ is J

Puzzle B:

$3x + 1 = y$ is G

$x = -\frac{y}{15} + 2$ is E

$-\frac{y}{4} + 3 = x$ is B

$-5x + 7 = y$ is F

$3x - 2y = 0$ is J

$y = 3x$ is C

$2y + 3x = 0$ is D

$x = \frac{(2y - 7)}{3}$ is K

$y = \frac{3}{4}x + 1$ is H

$-(5x - y) = -1$ is L

$-3x + 12 = y$ is A

$y = \frac{4}{3}x + 12$ is I

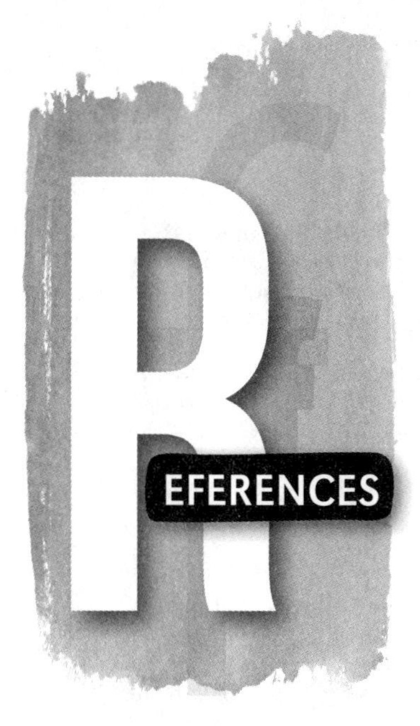

Beyraneyand, Matthew L. 2014. "Quick Reads: The Different Representations of Rational Numbers." *Mathematics Teaching in the Middle School* 19 (6): 382–385.

Boaler, Jo. 2013. "How to Learn Math." Stanford Online EDUC115N, July 14–September 28.

Bofferding, Laura. 2010. "Addition and Subtraction with Negatives: Acknowledging the Multiple Meanings of the Minus Sign." In *Proceedings of the 32nd Annual Conference of the North American Chapter of the International Group for the Psychology of Mathematics Education,* ed. Patricia Brosnan, Diana B. Erchick, and Lucia Flevares. Columbus, OH: The Ohio State University. http://pmena.org/2010/.

Bray, Wendy S. 2013. "How to Leverage the Potential of Mathematical Errors." *Teaching Children Mathematics* 19 (7): 424–431.

Bressoud, David. 2012. "Teaching and Learning for Transference." MAA Launchings. http://launchings.blogspot.com/2012/08/teaching-and-learning-for-transference.html.

Bureau of Labor Statistics, U.S. Department of Labor. *Occupational Outlook Handbook, 2014-15 Edition, Statisticians.* Also available online at http://www.bls.gov/ooh/math/statisticians.htm.

Chapin, Suzanne H., Catherine O'Connor, and Nancy C. Anderson. 2009. *Classroom Discussions: Using Math Talk to Help Students Learn.* 2nd ed. Sausalito, CA: Math Solutions.

Choppin, Jeffrey M., Carolyn B. Clancy, and Scott J. Koch. 2012. "Developing Formal Procedures through Sense-Making." *Mathematics Teaching in the Middle School* 17 (9): 252–257.

Cioe, Michael, Sherryl King, Deborah Ostien, Nancy Pansa, and Megan Staples. 2015. "Moving Students to 'the Why?'" *Mathematics Teaching in the Middle School* 20 (8): 484–491.

Cohen, Jessica S. 2013. "Strip Diagrams: Illuminating Proportions." *Mathematics Teaching in the Middle School* 18 (9): 536–542.

Collins, Anne, and Linda Dacey. 2010. *Zeroing in on Number and Operations: Key Ideas and Common Misconceptions, Grades 5–6.* Portland, ME: Stenhouse.

———. 2014. *It's All Relative: Key Ideas and Common Misconceptions About Ratio and Proportion, Grades 6–7.* Portland, ME: Stenhouse.

Common Core Standards Writing Team. 2011a. *Progressions for the Common Core State Standards in Mathematics: 6–7, Ratios and Proportional Relationships (draft).* Tucson, AZ: Institute for Mathematics and Education, University of Arizona. https://commoncoretools.files.wordpress.com/2012/02/ccss_progression_rp_67_2011_11_12_corrected.pdf.

Common Core Standards Writing Team. 2011b. *Progressions for the Common Core State Standards in Mathematics: 6-8, Statistics and Probability (draft)*. Tucson, AZ: Institute for Mathematics and Education, University of Arizona. http://commoncoretools.me/wpcontent/uploads/2011/12/ccss_progression_sp_68_2011_12_26_bis.pdf.

Consumer Reports. 2015. "No Basis for Comparison: Unit Pricing Can Be Misleading When Labels Are Inconsistent or Unclear." http://www.consumerreports.org/cro/magazine/2015/06/unit-pricing-can-be-confusing/index.htm

Cook-Sather, Alison. 2010. "Students as Learners and Teachers: Taking Responsibility, Transforming Education, and Redefining Accountability." *Curriculum Inquiry* 40 (4): 555–575.

Dunston, Pamela J., and Andrew M. Tyminski. 2013. "What's the Big Deal about Vocabulary?" *Mathematics Teaching in the Middle School* 19 (1): 38–45.

Dweck, Carol. 2007. *Mindset: The New Psychology of Success*. New York: Ballantine Books.

Erickson, Tim. 1989. *Get It Together: Math Problems for Groups, 4-12*. Berkeley, CA: Lawrence Hall of Science.

Farbermann, Boris L., and Ruzania G. Musina. 2004. "Picturing the Concepts: An Interactive Teaching Strategy." *Thinking Classroom* 5 (4): 12–16.

Gartland, Karen. 2014. *Integrating the Common Core in Mathematics for Grades 6-8*. Huntington Beach, CA: Shell Education.

Jackson, Kara J., Emily C. Shahan, Lynsey K. Gibbons, and Paul A Cobb. 2012. "Launching Complex Tasks." *Teaching Middle School Mathematics* 18 (1): 24–29.

Kazemi, Elham, and Allison Hintz. 2014. *Intentional Talk: How to Structure and Lead Productive Mathematical Discussions*. Portland, ME: Stenhouse.

Kliman, Marlene. 2006. "Math Out of School: Families' Math Game Playing at Home." *School Community Journal* 16 (2): 69–90.

Koellner, Karen, Mary Pittman, and Jonathan L. Brendefur. 2015. "Expect the Unexpected When Teaching Probability." *Mathematics Teacher* 245: 29–32.

Koster, Ralph. 2013. *A Theory of Fun for Game Design*. 2nd ed. Sebastopol, CA: O'Reilly Media.

Lamon, Susan. 2012. *Teaching Fractions and Ratios for Understanding*. 2nd ed. Mahwah, NJ: Erlbaum Associates.

Lan, Yu-Ju, Yao-Ting Sung, Ning-chun Tan, Chiu-Pin Lin, and Kuo-En Chang. 2010. "Mobile-Device-Supported Problem-Based Computational Estimation Instruction for Elementary School Students." *Educational Technology and Society* 13 (3): 55–69.

Lobato, Joanne, Amy Ellis, and Randall I. Charles. 2010. *Developing Essential Understanding of Ratios, Proportions, & Proportional Reasoning, Grades 6–8*. In *Developing Essential Understanding Series*, ed. Rose Mary Zbiek. Reston, VA: National Council of Teachers of Mathematics.

Michael, Joel. 2006. "Where's the Evidence That Active Learning Works?" *Advances in Physiology Education* 30 (4): 159–167.

National Governors Association Center for Best Practices (NGA) and Council of Chief State School Officers (CCSSO). 2010. *Common Core State Standards for Mathematics*. Washington, DC: NGA and CCSSO.

National Mathematics Advisory Panel (NMAP). 2008. *Foundations for Success: The Final Report of the National Mathematics Advisory Panel*. Washington, DC: U.S. Department of Education.

National Research Council. 2001. *Adding It Up: Helping Children Learn Mathematics*. Washington, DC: National Academy Press.

Newman, Rich. 2012. "Goal Setting to Achieve Results." *Leadership* 41 (3): 12–15, 16–18, 38.

Novotná, Jarmila, and Maureen Hoch. 2008. "How Structure Sense for Algebraic Expressions or Equations Is Related to Structure Sense for Abstract Algebra." *Mathematics Education Research Journal* 20 (2): 93–104.

Panasuk, Regina M. 2011. "Taxonomy for Assessing Conceptual Understanding in Algebra Using Multiple Representations." *College Student Journal* 45 (2). 219–232.

Park, Jungeun, Beste Güçler, and Raven McCrory. 2013. "Teaching Prospective Teachers about Fractions: Historical and Pedagogical Perspectives." *Educational Studies in Mathematics* 82 (3): 455–479. Also available online at http://dx.doi.org/10.1007/s10649-012-9440-8.

Polya, George. 1957. *How to Solve It: A New Aspect of Mathematical Method*. 2nd ed. New York: Doubleday.

Rathouz, Margaret, Nesrin Cengiz, Angela Krebs, and Rheta N. Rubenstein. 2014. "Tasks to Develop Language for Ratio Relationships." *Mathematics Teaching in the Middle School* 20 (1): 38–44.

Reeder, Stacey. 2012. "Take Off: Planes in Flight." *Mathematics Teaching in the Middle School* 17 (7): 403–408.

Schiro, Michael S. 2009. *Mega-fun Games and Puzzles for the Elementary Grades: Over 125 Activities That Teach Math Facts and Concepts*. San Francisco: Jossey-Bass.

Selmer, Sarah J., Johnna J. Bolyard, and James A. Rye. 2011. "Statistical Reasoning over Lunch." *Mathematics Teaching in the Middle School* 7 (5): 274–281.

Shay, Kathleen B. "Tracing middle school students' understanding of probability: A longitudinal study." PhD diss., Rutgers The State University of New Jersey - New Brunswick, 2008.

Small, Marian. 2013. *Eyes on Math*. New York: Teachers College Press, Reston, NJ: National Council of Teachers of Mathematics.

Star, Jon R., and Bethany Rittle-Johnson. 2009. "Making Algebra Work: Instructional Strategies That Deepen Student Understanding, Within and Between Representations." *ERS Spectrum* 27 (2): 11–18.

Steen, Lynn A. 2004. *Achieving Quantitative Literacy*. Washington, D.C.: The Mathematical Association of America.

Tassell, Janet, Rebecca Stobaugh, and Linda Sheffield. 2011. "Developing Middle Grade Students' M3: Mathematical Promise, Passion, and Perseverance." *Parenting for High School* 1 (2): 6–9.

Van de Walle, John A., Karen S. Karp, and Jennifer M. Bay-Williams. 2013. *Elementary and Middle School Mathematics: Teaching Developmentally*. 8th ed. New York: Pearson Education.

Van de Walle, John A., Jennifer M. Bay-Williams, Lou Ann H. Lovin, and Karen S. Karp. 2013. *Student-Centered Mathematics: Developmentally Appropriate Instruction for Grades 6–8*. 2nd ed. New York: Pearson Education.

Warshauer, Hiroko. 2015. "Strategies to Support Productive Struggle." *Mathematics Teaching in the Middle School* 20 (7): 390–393.

Wilburne, Jane M., Tara Wildmann, Michael Morret, and Julie Stipanovic. 2014. "Classroom Strategies to Make Sense and Persevere." *Mathematics Teaching in the Middle School* 20 (3): 144–151.